Smart Innovation, Systems and Technologies

Volume 85

About this Series

The Smart Innovation, Systems and Technologies book series encompasses the topics of knowledge, intelligence, innovation and sustainability. The aim of the series is to make available a platform for the publication of books on all aspects of single and multi-disciplinary research on these themes in order to make the latest results available in a readily-accessible form. Volumes on interdisciplinary research combining two or more of these areas is particularly sought.

The series covers systems and paradigms that employ knowledge and intelligence in a broad sense. Its scope is systems having embedded knowledge and intelligence, which may be applied to the solution of world problems in industry, the environment and the community. It also focusses on the knowledge-transfer methodologies and innovation strategies employed to make this happen effectively. The combination of intelligent systems tools and a broad range of applications introduces a need for a synergy of disciplines from science, technology, business and the humanities. The series will include conference proceedings, edited collections, monographs, handbooks, reference books, and other relevant types of book in areas of science and technology where smart systems and technologies can offer innovative solutions.

High quality content is an essential feature for all book proposals accepted for the series. It is expected that editors of all accepted volumes will ensure that contributions are subjected to an appropriate level of reviewing process and adhere to KES quality principles.

More information about this series at http://www.springer.com/series/8767

Ioannis Hatzilygeroudis · Vasile Palade
Editors

Advances in Hybridization of Intelligent Methods

Models, Systems and Applications

Editors
Ioannis Hatzilygeroudis
School of Engineering, Department of Computer Engineering and Informatics
University of Patras
Patras
Greece

Vasile Palade
The Faculty of Engineering, Environment and Computing, School of Computing
Coventry University
Coventry
UK

ISSN 2190-3018 ISSN 2190-3026 (electronic)
Smart Innovation, Systems and Technologies
ISBN 978-3-319-88320-5 ISBN 978-3-319-66790-4 (eBook)
https://doi.org/10.1007/978-3-319-66790-4

Printed on acid-free paper

This Springer imprint is published by Springer Nature
The registered company is Springer International Publishing AG
The registered company address is: Gewerbestrasse 11, 6330 Cham, Switzerland

Preface

The invention of hybrid intelligent methods is a very active research area in artificial intelligence (AI). The aim is to create hybrid methods that benefit from each of their components. It is generally believed that complex problems can be easily solved with hybrid methods. By "hybrid," we mean any kind of combined use (either tight or loose) of distinct intelligent methods toward solving a problem, either specific or general. In this sense, it is used as synonymous with "integrated."

Some of the existing efforts try to make hybrids of what are called soft computing methods (fuzzy logic, neural networks, and genetic algorithms) either among themselves or with more traditional AI methods, such as logic and rules. Another stream of efforts integrates case-based reasoning or machine learning with soft computing or traditional AI methods. Yet another integrates agent-based approaches with logic and non-symbolic approaches. Some of the combinations have been quite important and more extensively used, like neuro-symbolic methods, neuro-fuzzy methods, and methods combining rule-based and case-based reasoning. However, there are other combinations that are still under investigation, such as those related to the Semantic Web and Big Data areas. For example, the recently emerged deep learning architectures or methods are also hybrid by nature. In some cases, integrations are based on first principles, creating hybrid models, whereas in other cases they are created in the context of solving problems leading to systems or applications.

Important topics of the above area are (but not limited to) the following:

- Case-Based Reasoning Integrations
- Ensemble Learning, Ensemble Methods
- Evolutionary Algorithms Integrations
- Evolutionary Neural Systems
- Fuzzy-Evolutionary Systems
- Semantic Web Technologies Integrations
- Hybrid Approaches for the Web
- Hybrid Knowledge Representation Approaches/Systems
- Hybrid and Distributed Ontologies

- Information Fusion Techniques for Hybrid Intelligent Systems
- Integrations of Neural Networks
- Integrations of Statistical and Symbolic AI Approaches
- Intelligent Agents Integrations
- Machine Learning Combinations
- Neuro-Fuzzy Approaches/Systems
- Swarm Intelligence Methods Integrations
- Applications of Combinations of Intelligent Methods to
  - Biology and Bioinformatics
  - Education and Distance Learning
  - Medicine and Health Care
  - Multimodal Human–Computer Interaction
  - Natural Language Processing and Understanding
  - Planning, Scheduling, Search, and Optimization
  - Robotics
  - Social Networks

This volume includes extended and revised versions of some of the papers presented in the 6th International Workshop on Combinations of Intelligent Methods and Applications (CIMA 2016) and also papers submitted especially for this volume after a CFP. CIMA 2016 was held in conjunction with the 22nd European Conference on Artificial Intelligence (ECAI 2016), August 30, 2016, The Hague, Holland. Papers went through a peer review process by the CIMA-16 program committee members.

Giannopoulos et al. present results on using two deep learning methods (GoogleNet and AlexNet) on facial expression recognition. The paper of Haque et al. presents results on how communication model affects robotics swarm performance. Jabreel et al. introduce and experiment with a target-dependent sentiment analysis approach for tweets. Maniak et al. present a hybrid approach used for the modeling and prediction of taxi usage in the context of smart cities. The paper of Mason et al. introduces a reinforcement learning approach combining a Markov decision process and quantification verification to restrict an agent's behavior at a safe level. Mosca and Magoulas propose a method for approximating an ensemble of deep neural networks by a single deep neural network. Finally, Teppan and Friedrich present a constraint answer programming solver and investigate its performance through its application to two manufacturing problems.

We would like to express our appreciation to all the authors of submitted papers as well as to the members of CIMA 2016 program committee for their excellent review work.

We hope that this kind of post-proceedings will be useful to both researchers and developers.

Patras, Greece Ioannis Hatzilygeroudis
Coventry, UK Vasile Palade

Reviewers (From CIMA 2016 Program Committee)

Plamen Agelov, Lancaster University, UK
Nick Bassiliades, Aristotle University of Thessaloniki, Greece
Kit Yan Chan, Curtin University, Australia
Gloria Cerasela Crisan, Vasile Alecsandri University of Bacau, Romania
Georgios Dounias, University of the Aegean, Greece
Foteini Grivokostopoulou, University of Patras, Greece
Ioannis Hatzilygeroudis, University of Patras, Greece (Co-chair)
Andreas Holzinger, TU Graz and MedUni Graz, Austria
George Magoulas, Birkbeck College, UK
Christos Makris, University of Patras, Greece
Antonio Moreno, University Rovira i Virgili, Spain
Vasile Palade, Coventry University, UK (Co-chair)
Isidoros Perikos, University of Patras, Greece
Camelia Pintea, Technical University of Cluj-Napoca, Romania
Jim Prentzas, Democritus University of Thrace, Greece
Roozbeh Razavi-Far, Politecnico di Milano, Italy
David Sanchez, University Rovira i Virgili, Spain
Kyriakos Sgarbas, University Of Patras, Greece
Douglas Vieira, Enacom-Handcrafted Technologies, Brazil

Contents

Deep Learning Approaches for Facial Emotion Recognition: A Case Study on FER-2013

Panagiotis Giannopoulos, Isidoros Perikos and Ioannis Hatzilygeroudis

Abstract Emotions constitute an innate and important aspect of human behavior that colors the way of human communication. The accurate analysis and interpretation of the emotional content of human facial expressions is essential for the deeper understanding of human behavior. Although a human can detect and interpret faces and facial expressions naturally, with little or no effort, accurate and robust facial expression recognition by computer systems is still a great challenge. The analysis of human face characteristics and the recognition of its emotional states are considered to be very challenging and difficult tasks. The main difficulties come from the non-uniform nature of human face and variations in conditions such as lighting, shadows, facial pose and orientation. Deep learning approaches have been examined as a stream of methods to achieve robustness and provide the necessary scalability on new type of data. In this work, we examine the performance of two known deep learning approaches (GoogLeNet and AlexNet) on facial expression recognition, more specifically the recognition of the existence of emotional content, and on the recognition of the exact emotional content of facial expressions. The results collected from the study are quite interesting.

Keywords Affective computing · Facial emotion recognition · Deep learning · Convolutional neural networks · GoogLeNet · Alexnet

P. Giannopoulos (✉) · I. Perikos · I. Hatzilygeroudis
Department of Computer Engineering and Informatics,
University of Patras, Patras, Greece
e-mail: giannopp@ceid.upatras.gr

I. Perikos
e-mail: perikos@ceid.upatras.gr

I. Hatzilygeroudis
e-mail: ihatz@ceid.upatras.gr

I. Perikos
Technological Educational Institute of Western Greece, Patras, Greece

I. Hatzilygeroudis and V. Palade (eds.), *Advances in Hybridization of Intelligent Methods*, Smart Innovation, Systems and Technologies 85,
https://doi.org/10.1007/978-3-319-66790-4_1

1 Introduction

An envisaged aim of artificial intelligence is to make the interaction between human and next generation computing systems more natural. In order to achieve efficient and smooth interaction between human and computer systems, a series of aspects of human behavior should be taken into account. One of the most important aspects concerns the emotional behavior and the affective state of the human. Next generation human-centered computing systems should possess the capacity to perceive, accurately analyze and deeply understand emotions as communicated by social and affective channels [1].

Emotions constitute an innate and important aspect of human behavior that colors the way of communication. Humans express their innate conditions through various channels, such as body language and facial expressions. Facial expressions are the most direct and meaningful channel of non-verbal communication, which forms a universal language of emotions that can instantly express a wide range of human emotional states, feelings and attitudes and assists in various cognitive tasks. The accurate analysis and interpretation of the emotional content of human facial expressions is essential for the deeper understanding of human behavior. Indeed, facial expressions are to wit the most cogent, naturally preeminent means for human beings to communicate emotions, comprehension, and intentions and to regulate interactions and communication with other people [2, 1].

Facial expressions considerably assist in direct communication and it has been indicated that during face-to-face human communication, 7% of the information is communicated by the linguistic part, such as the spoken words, 38% is communicated by paralanguage, such as the vocal part, and 55% is communicated by the facial expressions [3]. Indeed, even a simple signal such as a head nod or a smile can convey a large number of meanings [4, 5]. In general, facial expressions are the most natural, meaningful and important communication channel of human interaction and communication.

The recognition of facial expressions is assistive in a wide spectrum of systems and applications and is quite necessary for achieving naturalistic interaction. The facial expressions assist in various cognitive tasks; so reading and interpreting the emotional content of human expressions is essential for deeper understanding of human condition. Therefore, the main aim of facial expression recognition methods and approaches is to enable machines to automatically estimate the emotional content of a human face. Giving computer applications the ability to recognize the emotional state of humans from their facial expressions is a very important and challenging task with wide ranging applications. In general, affective computing systems need to perceive emotional reactions by the user and successfully incorporate this information into the interaction process [6]. The interaction between human and computer systems (HCI) would become much more natural and vivid if the computer applications could recognize and adapt to the emotional state of the human. Indeed, automated systems that can determine emotions of humans based on their facial expressions can improve the human computer interaction and give computer systems the ability to customize and adapt their responses [7]. Embodied conversational agents can greatly benefit from spotting and understanding the emotional states of the participants, achieving more realistic interactions at an emotional level [8]. In intelligent tutoring systems, emotions and learning are inextricably bound together; so recognizing learners' emotional states could significantly improve the efficiency of the learning procedures delivered to them [9–11]. Moreover, surveillance applications such as driver monitoring and elderly monitoring systems could benefit from a facial emotion recognition system, gaining the ability to deeper understand and adapt to the person's cognitive and emotional condition. Also, facial emotion recognition could be applied to medical treatment to monitor patients and detect their status.

However, the analysis of human face characteristics and the recognition of its emotional state are considered to be very challenging and difficult tasks. The main difficulty comes from the non-uniform nature of the human face and various limitations related to lighting, shadows, facial pose and orientation conditions [12]. Although a human can detect and interpret faces and facial expressions naturally, with little or no effort, accurate and robust facial expression recognition by computer systems is still a great challenge. Deep learning approaches have been examined as a stream of methods to achieve robustness and superior performance compared to basic machine learning classification methods, such as multilayer perceptron neural networks and support vector machines. Since humans operate in diversity of contexts, human behavior analysis needs to be robust and deep learning methods can provide the necessary robustness and scalability on new type of data.

In this work, we examine the performance of deep learning approaches on facial expression recognition, more specifically on the recognition of the existence of emotional content on facial expressions and on the recognition of the exact emotional content. To that end, we experiment with GoogLeNet and Alexnet, two popular and wide used deep learning methods, and we examine their performance on FER-2013 dataset. The results collected from the examination study are quite interesting.

The structure of the paper is organized as follows. In Sect. 2, background topics on emotion models and on deep learning methods are examined. In Sect. 3, related works on the utilization of machine learning and deep learning methods for the recognition of facial expressions are presented. After that, Sect. 4, presents the experimental study conducted, examines the results collected and discusses the main findings. Finally, Sect. 5 concludes the paper and draws directions for future work.

2 Background Topics

2.1 *Facial Emotion Recognition Methods*

In the field of facial emotion recognition two types of methods dominate: the holistic methods and the analytical or local-based methods [13]. The holistic methods try to model the human facial deformations globally, which encode the entire face as a whole. On the other hand, the analytical methods observe and measure local or distinctive human facial deformations such as eyes, eyebrows, nose, mouth etc. and their geometrical relationships in order to create descriptive and expressive models [14]. In the feature extraction process for expression analysis there are mainly two types of approaches, which are the geometric feature based methods and the appearance based methods. The geometric facial features try to represent the geometrical characteristics of a facial part deformation, such as the part's locations, and model its shape. The appearance based methods utilize image filters, such as Gabon wavelets, to the whole face or on specific parts to extract feature vectors.

2.2 *Emotion Models*

The way that emotions are represented is a basic aspect of an emotion recognition system. A very popular categorical model is the Ekman emotion model [15], which specifies six basic human emotions: anger, disgust, fear, happiness, sadness, surprise. Ekman's emotion model has been used in several research studies and in various systems that are used to recognize emotional state from text and facial expressions. Another popular model is the OCC (Ortony/Clore/Collins) model [16], which specifies 22 emotion categories based on emotional reactions to situations and is mainly designed to model human emotions in general. Plutchik's model of emotions [17] is a dimensional model, which offers an integrative theory based on evolutionary principles and defines eight basic bipolar emotions. These eight emotions are organized into four bipolar sets: joy versus sadness, anger versus fear, trust versus disgust, and surprise versus anticipation.

2.3 *Deep Learning Methods*

Deep learning methods constitute a stream of approaches that rely on deep architectures and are surpassing other machine learning approaches in terms of accuracy and efficiency. In general, deep architectures are composed of multiple levels of non-linear operations components, like neural nets that have many hidden layers. They aim to learn feature hierarchies with features at higher levels in the hierarchy formed by the composition of lower level features. Deep learning methods through machine learning models with multi hidden layers, which are trained on massive volumes of data, could learn more useful features and thus improve the accuracy of classification and prediction [18]. In general, they employ the softmax activation function for prediction and minimize cross-entropy loss. Deep learning methods are becoming exponentially more important, due to their demonstrated success at tackling complex learning problems. Some widely used methods among others are the Convolutional Neural Networks, the Deep Belief Networks and the Deep Boltzmann Machines.

Convolutional Neural Networks.

Convolutional neural networks (CNNs) are a category of neural networks specialized in areas such as image recognition and classification. In general, convolutional neural networks pose collections of small neurons in multiple layers that process the input image in portions that are the receptive fields. A convolutional neural network, in the most part of it, consists of three layer types: the convolutional layers, the max-pooling layers and the fully-connected layer. With the latter one being the less fundamental, the other two types of layers are responsible for feature extraction, introduction of non-linearity in the network and feature dimension reduction. The fully-connected layer is assigned the task of classifying the input, based on the previously extracted features by the other layers. The training process of a convolutional neural network is called back propagation and it can be divided into 4 parts, the forward pass, the loss function, the backward pass and the weights update. A simple architecture of a CNN is shown in (Fig. 1) and was created by [44].

Deep Belief Networks.

Deep belief networks (DBNs) are multi-layer belief networks that are probabilistic generative models, which resemble convolutional neural networks in many ways. Deep belief networks composed of multiple layers of latent variables hidden units, where there are connections between the layers, but not between the units within each layer. The main problem of back propagation used in CNNs is the possibility of hitting a local minima instead of the global one, when performing the gradient descent. Unlike convolutional neural networks, deep belief networks are not trained

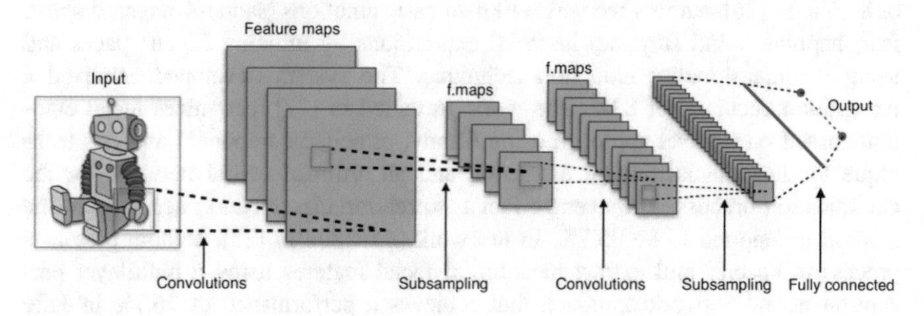

Fig. 1 Representation of basic convolutional neural network architecture

via the back propagation process immediately. Instead, the "pre-training" process takes place and through i, the error rate accosts the optimal. Then, the back propagation process commits to reduce the error rate even more. The reasons that enable pre-training to be of any success are better initialization of the weights and soft regularization, which offers better generalization power for the back propagation process later on. The whole deep belief network is trained when the learning for the final hidden layer is achieved.

Deep Boltzmann Machines.

Deep Boltzmann Machines (DBM's) [19] relay on Boltzmann machines, which are networks of symmetrically connected neuron-like units that make stochastic decisions about whether to be on or off. Deep Boltzmann Machines can be considered a special Boltzmann machine where the hidden units are organized in deep layered manners, where the adjacent layers are connected and there are no visible hidden connections within the same layer [20]. Deep Boltzmann Machines have the potential of learning internal representations that become increasingly complex, something that is considered to be a promising approach for problems like facial expression recognition [21].

3 Related Work

Over the last decade there has been a huge research interest and many studies on the formulation of methods and systems for the recognition of emotional content of facial expressions. A detailed overview of approaches can be found in [22, 23]. Several works study the way humans express emotions and try to specify facial emotions from static images and video streams [24].

The work in [25], presents a facial expression classification method based on histogram sequence of feature vector. It consists of four main tasks, which are image pre-processing, mouth segmentation, feature extraction and classification, which is based on histogram-based methods. The system is able to recognize five human expressions: happiness, anger, sadness, surprise and neutral, based on the geometrical characteristics of the human mouth with an average recognition accuracy of 81.6%. In [26], authors recognize Ekman basic emotions (sadness, anger, disgust, fear, happiness and surprise) in facial expressions by utilizing Eigen spaces and using a dimensionality reduction technique. The system developed achieved a recognition accuracy of 83%. The work presented in [27] recognizes facial emotions based on a novel approach using Canny, principal component analysis technique for local facial feature extraction and an artificial neural network for the classification process. The average facial expression classification accuracy of the method is reported to be 85.7%. In the work presented in [28], authors present a process to specify and extract meaningful facial features using a multilayer perceptron neural network approach that achieves a performance of 76.7% in Jaffe database. In [29], a hybrid two stage classification schema is presented, where a

SVM is used to specify whether facial expressions convey emotional content or are neutral and then, at the second stage, a Multilayer Perceptron Neural Network specifies each expression's emotional content on Ekman's emotional categories. The system reports a performance of 85% on Jaffe and Cohn-Kanade database. In the work presented in [30], authors developed a system that consists of a face tracker, an optical flow algorithm to track motion in faces and a recognition engine based on SVMs and multilayer perceptron, which achieve a recognition accuracy of 81.8%. The authors in the work presented in [31] recognize four basic emotions of happiness, anger, surprise and sadness focusing in preprocessing techniques for feature extraction, such as Gabor filters, linear discrimination analysis and principal component analysis. They achieve a 93.8% average accuracy in their experiments, for images of the Jaffe face database with little noise and with particularly exaggerated expressions and an average accuracy of 79% in recognition on just smiling/non smiling expressions in the ORL database. In [32], a work that recognizes the seven emotions on Jaffe database using Fisher weight map is presented. Authors utilize image preprocessing techniques such as illumination correction and histogram equalization and the recognition rate of their approach is reported to be 69.7%. Principal component analysis (PCA) and Linear discriminant analysis (LDA) methods are used for both dimensional reduction and also the expression classification in emotion recognition process [33]. In the work presented in [34], authors highlight the higher performance of PCA-LDA fusion methods. In [35], representation-based classification of facial expressions. The PCA-based dictionary building results in better recognition performance on CK +, MMI databases up to 6%.

Recently deep learning methods have attracted a lot of interest and there is a huge research interest in the study of deep learning methods on the recognition of emotions from facial expressions. Deep Learning methods have been examined on facial emotion recognition and are mainly based on supervised learning relying on manually labelled data. In the work presented in [36], authors present a L2-SVM, which has as its main objective to train deep neural net. Lower layer weights are learned by back propagating the gradients from the top layer linear SVM. The authors approach achieved an accuracy of 71.2% on FER dataset. Authors, in the work presented in [37], present a bag of visual words model adapted to the FER Challenge dataset, where histograms of visual words were replaced with normalized presence vectors, then local learning was used to predict class labels of test images. Their model reported an accuracy of 67.49% on FER dataset. In the work presented in [38], a deep network that consists of two convolutional layers, each followed by max pooling and then four Inception layers, is presented. The network is a single component architecture that takes registered facial images as the input and classifies them into either the six basic expressions or the neutral expression. The author's approach reports an accuracy of 66.4% on FER dataset.

4 Experimental Study

In this section, the experimental study conducted and the results gathered are presented. The dataset used in the experiment is the FER-2013 dataset, a widely used dataset for benchmark and assessing the performance of facial expression recognition systems and approaches.

4.1 Datasets

The Facial Emotion Recognition 2013 (FER-2013) dataset was created by Pierre Luc Carrier and Aaron Courville and was introduced in the ICML 2013 workshop's facial expression recognition challenge [39]. The dataset was formulated by using the Google image search API to crawl images that match emotional keywords. In total, the dataset consists of 35887 facial images most of them in wild settings. In the dataset, 4953 images express anger, 547 images express disgust, 5121 images express fear, 8989 images express happiness, 6077 images express sadness, 4002 images express surprise and 6198 images are emotionally neutral. The dataset is quite challenging, since faces greatly vary in age, pose and occlusion conditions. In addition, the accuracy of human recognition is approximately 65 ± 5%. The dataset comprises 3 parts: the original training data (OTD), which consists of 28709 images, the Public Test Data (PTD), which consists of 3589 images and was used during the workshop contest to provide feedback accuracy of participating models, and the Final Test Data (FTD), which consists of 3589 images that were used to score the final models (Fig. 2).

Fig. 2 Example images from the FER-2013 dataset [45]

4.2 Deep Learning Architectures

GoogLeNet

The incarnation of the "Inception" architecture proposed in the ILSVRC 2014 competition was named GoogLeNet [40]. The network is 22 layers deep, when counting only layers with parameters, while the overall number of layers used is almost 100. From Fig. 3, it is obvious that there are parts of the network which are executed in parallel. These parts are called "Inception modules". The initial naïve approach was to split the input of each layer to 1 × 1, 3 × 3 and 5 × 5 convolutions and a 3 × 3 pooling, but that would lead to an extortionate number of outputs. Shifting from the naïve to the full Inception module, the creators decided to combat this by inserting 1 × 1 convolution operations before the 3 × 3 and 5 × 5 ones. By adding the 1 × 1 convolutions, a kind of dimensional reduction is introduced and the problem is significantly mitigated. Nine such modules are used in the network, until we reach the last layers, where the classification process takes place. Contrary to most deep learning architectures, the GoogLeNet does not use fully connected layers. Instead, it uses an average pool to transfer from a 7 × 7 × 1024 to a 1 × 1 × 1024 volume, using far more less parameters. For the purposes of comparison with the AlexNet architecture, which will be deployed next, GoogLeNet uses 12 times fewer parameters. Another main idea of the proposed architecture is to make use of the extra sparsity, but exploiting the hardware, by utilizing computations on dense matrices. Clustering sparse matrices into relatively dense submatrices is what gives state of the art results. The network was trained through the DistBelief system [41], using asynchronous stochastic gradient descent of 0.9 momentum, fixed learning rate schedule (decreasing the learning rate by 4% every 8 epochs), Polyak averaging at inference time, photometric distortions to combat overfitting, and random interpolation methods for resizing. The performance is measured based on the highest scoring classifier predictions and the main metrics are: top-1 accuracy rate (compares the ground truth against the first predicted class) and top-5 error rate (compares the ground truth against the first 5 predicted classes).

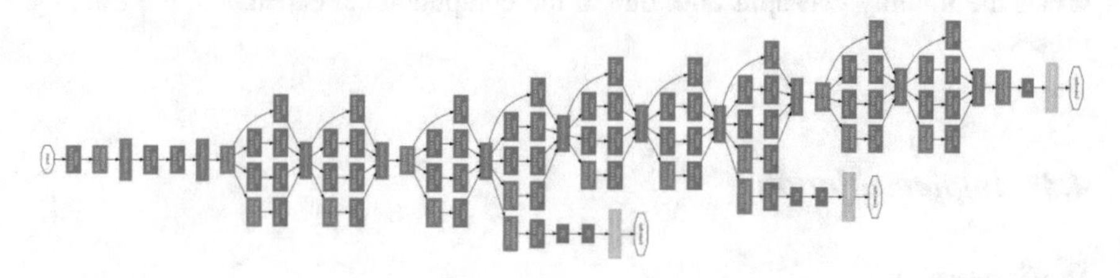

Fig. 3 Basic architecture of GoogLeNet

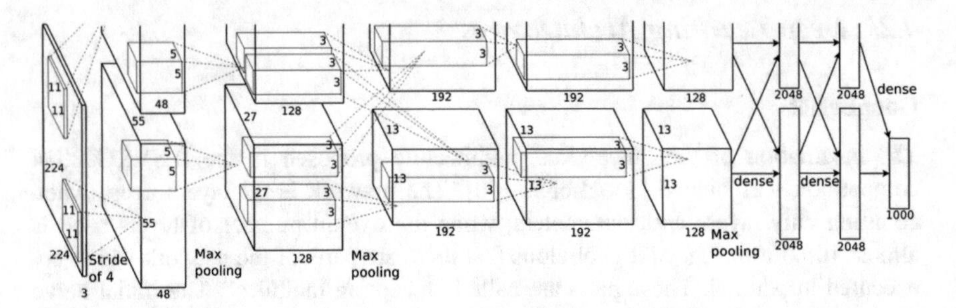

Fig. 4 Basic architecture of AlexNet

AlexNet.

AlexNet, named by its creator Alex Krizhevsky, is an architecture originally submitted in the ILSVRC competition in 2012 [42]. The network consists of 8 fully-connected. The output of the last fully-connected layer is fed to a 1000-way softmax function, which produces a distribution over 1000 class labels. Starting with the first convolutional layer shown in Fig. 4, it is clear that the input size is $224 \times 224 \times 3$, the receptive field size is equal to 11, the stride is 4, and the output of the layer is $55 \times 55 \times 96$, as there are two streams of depth 48 each. This means that there are $55 \times 55 \times 96 = 290400$ neurons, each one having $11 \times 11 \times 3 = 363$ weights and 1 bias, leading to $290400 \times 364 = 105705600$ parameters on the first layer. This makes overfitting inevitable when it comes to learning all of them. Two approaches on handling over-fitting were proposed on [42] (data augmentation and dropout), but we would branch off the goal of this case study by delving into them. Keeping up with Fig. 4, we see that the second layer has 256 kernels of size $5 \times 5 \times 48$, the next one has 384 kernels of size $3 \times 3 \times 256$ etc. The network was trained using stochastic gradient descent with a batch size of 128, momentum of 0.9 and weight decay of 0.0005. The weights in each layer were initialized from a zero-mean Gaussian distribution with a standard deviation of 0.01. Finally, equal learning rate for all layers was used, and when the validation error rate stopped improving, the learning rate was divided by 10. The architecture, as shown in Fig. 4, is divided into two streams, which are the two GPUs the training was split into, due to the computational expense of the training process.

4.3 Implementation

Pre-Processing.

The FER-2013 dataset holds 48×48 pixel values for each image. Initially, we rebuild the images of the dataset via Python scripts that receive as input the intensity of the black color in each image, meaning a vector of the 48×48 values mentioned above, with every value being between 0 and 255. This produces two folders with the train and test images. Both of those folders, combined with the text files that describe the label of each image, which were also created simultaneously by the same Python script we crafted, are used in order to create the "lmdb" files needed by the framework.

Implementation Framework.

The framework used for the experimental purpose of this case study is the "Caffe" toolbox [43], which is maintained by Berkley Vision and Learning Center (BVLC) and has acquired many contributors from its growing community. Caffe toolbox was chosen mainly for the high maturity level it encompasses and also the suitability it offers for using, manipulating and training deep learning methods as well as the GoogLeNet and AlexNet, described in the previous sections.

Training process and hyper parameter values.

The training process was designed to go through 5000 iterations on the GoogLeNet and the AlexNet experiments. The average loss was presented every 10 iterations, while the accuracy of each network was presented every 500 iterations. The values and methods mentioned below were applied to both architectures. We fed the *source* field of the *train_val.prototxt* file with the path to the lmdb files we created, as this is the format of input files that Caffe recognizes, we set the *batch_size* of the train layer to 32 and the *batch_size* of the test layer to 50, *test*_iter to 20, all *mean_value* fields to 125, and *crop_size* as both architectures require it to be; 224. As for the most crucial part, the hyper parameters were set as; *average_loss* = 40, *base_lr* = 0.001, *lr_policy* = poly, *power* = 0.5, *max_iter* = 5000, *momentum* = 0.9, *weight_decay* = 0.0002, *test_interval* = 500 and finally, the *solver_mode* = CPU. The weights from the pretrained BVLC models were used as initials. Various alternations of our final approach were studied, for the purpose of ensuring that the best optimal, to our knowledge, hyper parameter values were chosen. We next describe the most notable. In an attempt to expand the network's learning capability, we adjusted the learning rate to 0.0001 and set the max iterations to 10000. Despite the promising results we received at the first iterations, the accuracy seemed to approach the one we reached using the learning rate and max iterations mentioned above. Thus, presenting both cases would be of no use. We then attempted to train the network on more iterations with the initial learning rate of 0.001, and found out that after the first 5000 iterations, the accuracy was almost the same. Moreover, as more iterations were performed, the accuracy started to slightly drop, meaning the network was overtraining. Another endeavor was to increase the number of test iterations performed, from 20 to 100, aiming to make the testing process more robust and perhaps raising its accuracy. When using a *test_iter* of μ for example, the network is performing μ*batch_size (of the test layer) tests on individual inputs, calculating the probabilities and classifying each input to the most relatable class. Again, the accuracy did not prove to augment noticeably and *test_iter* was kept as equal to 20.

4.4 Performance Analysis

The performance of the GoogLeNet and AlexNet has been studied on the FER-2013 dataset in three aspects, where each one evaluates a specific functionality of the methods. In the first part of the study, the performance of the networks has been studied in recognizing the existence, or not, of emotional content in a facial expression and after that, in the second part of the study, their performance has been examined in specifying the exact emotional content of facial expressions. Finally, in the third part the two deep learning methods have been trained on both emotional and neutral data studied. The results of the three parts are illustrated in Fig. 5.

Recognizing the existence of emotional content.
In this section of the study, the dataset was divided into two parts. The first part contained the neutral emotion (represented by class "6" based on the FER-2013 documentation), and the second part contained the rest of the classes (from "0" to "5"), which represent any emotional state. In this specific binary classification, the two networks start the testing process after the first 500 training iterations with exactly the same accuracy. Through the 2500 first iterations it is obvious that the deviation between them is minor. AlexNet only seems to surpass GoogLenet's accuracy when 4000 iterations are completed, by 1%, and then again GoogLeNet finishes the 5000 iterations with a greater accuracy. Moreover, by observing Fig. 5, it is clear that both network accuracies deploy almost parallel to each other after every testing phase, with the most obvious part of the parallel formation being the last 500 iterations. No major divergences appear in the specific experiment, as there is not a wide variety of features to be distinguished, since only two classes exist.

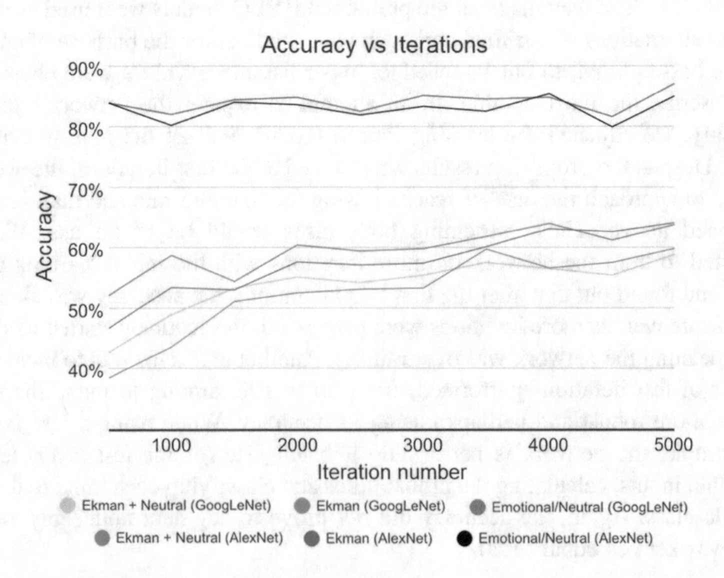

Fig. 5 The performance based on the concept studied

Recognizing the emotional content of emotional expressions.

In this phase of the study, the neutral emotion was removed from the dataset and the computer was trained in recognizing emotions and the class they belong to. Unlike the previous experiment, here, GoogLeNet seems to begin the testing process after the first 500 training iterations with an accuracy of 8% higher than AlexNet. The two accuracies approach on 1500 iterations, and then they seem to deploy almost parallel to each other, until they reach the 4500th iteration where AlexNet is leading with a 1.4% higher rate than GoogLeNet. Despite the initial distance, AlexNet managed to complete the training process, achieving almost the same accuracy. The reason for this is that the GoogLeNet architecture is much deeper, offering a wider range of feature recognition. On the other hand, a deeper network leads to a slower training process, and this is why AlexNet seems to almost achieve GoogLeNet's accuracy at the end of 5000 iterations.

Recognizing the emotional (emotional/neutral) content of emotional expressions.

Finally, in the last phase of the study, the whole dataset was used for the training process. This means that the computer had to recognize the existence of 7 different emotions, assuming that the neutral expression represents a separate emotional class. This is the only approach of the study that AlexNet seems to be more accurate than GoogLeNet in almost half of the iterations. Again, AlexNet begins at a much lower accuracy and after the first 3000 iterations, it surpasses the accuracy of GoogLeNet and manages to preserve this lead until the end. Again both architectures end the training and testing processes with almost the same accuracy, but this time AlexNet is ahead by 1%. By using the whole dataset as mentioned, an extra complexity of feature recognition is introduced, and GoogLeNet might be more efficient after many iterations when AlexNet will be overtrained, due to its lack of depth, but in the first 5000 iterations, thanks to its shallow architecture, AlexNet seems to reach and get ahead of GoogLeNet's accuracy as it manages to train quite quicker.

5 Conclusions and Future Work

The accurate analysis and interpretation of the emotional content of human facial expressions is essential for deeper understanding human behavior. Although a human can detect and interpret faces and facial expressions naturally, with little or no effort, accurate and robust facial expression recognition by computer systems is still a great challenge. Deep learning approaches have been examined as a stream of methods to achieve robustness and provide the necessary robustness and scalability on new type of data. In this work, we examine the performance of deep learning methods on facial expression recognition and more specifically, the recognition of the existence of emotional content and the recognition of the exact emotional content of the facial expressions. The experimental study examined the performance of GoogLenet and Alexnet on the FER-2013 dataset and revealed quite interesting results. As future work, a main direction concerns the design and the examination of the performance of ensembles of deep learning methods. Also another direction for future work concerns the examination of image preprocessing procedures and oration alignments processes and how they affect the performance of the deep learning methods. Exploring this direction, is a main aspect that future work will examine.

References

1. Pantic, M.: Facial expression recognition. In: Encyclopedia of Biometrics, pp. 400–406. Springer, US (2009)
2. Ekman, P., Rosenberg, E.L. (eds.): What the face Reveals: Basic and Applied Studies of Spontaneous Expression Using the Facial Action Coding System. Oxford University Press, Oxford, UK (2005)
3. Mehrabian, A.: Communication without words. Psychol. Today **2**(4), 53–56 (1968)
4. Heylen, D.: Head gestures, gaze and the principles of conversational structure. Int. J. Humanoid Rob. **3**(03), 241–267 (2006)
5. Ochs, M., Niewiadomski, R., Pelachaud, C.: Facial Expressions of Emotions for Virtual Characters. The Oxford Handbook of Affective Computing, **261** (2014)
6. Liebold, B., Richter, R., Teichmann, M., Hamker, F.H., Ohler, P.: Human capacities for emotion recognition and their implications for computer vision. *i-com*, **14**(2), pp. 126–137 (2015)
7. Clavel, C.: Surprise and human-agent interactions. Rev. Cogn. Linguist. **13**(2), 461–477 (2015)
8. Liebold, B., Ohler, P.: Multimodal emotion expressions of virtual agents, mimic and vocal emotion expressions and their effects on emotion recognition. In: Humaine Association Conference on Affective Computing and Intelligent Interaction (ACII), pp. 405–410. IEEE (2013)
9. Bahreini, K., Nadolski, R., Westera, W.: Towards multimodal emotion recognition in e-learning environments. Interact. Learning Environ. **24**(3), 590–605 (2016)
10. Akputu, K.O., Seng, K.P., Lee, Y.L.: Facial emotion recognition for intelligent tutoring environment. In: 2nd International Conference on Machine Learning and Computer Science (IMLCS'2013), pp. 9–13 (2013)
11. Shen, L., Wang, M., Shen, R.: Affective e—learning: Using "emotional" data to improve learning in pervasive learning environment. Educ. Technol. Soc. **12**(2), 176–189 (2009)
12. Koutlas, A., Fotiadis, D.I.: An automatic region based methodology for facial expression recognition. In: IEEE International Conference on Systems Man and Cybernetics SMC, pp. 662–666 (2008)
13. Pantic, M., Rothkrantz, L.J.M.: Automatic analysis of facial expressions: The state of the art. Pattern Anal. Mach. Intell. IEEE Trans. **22**(12), 1424–1445 (2000)
14. Arca, S., Campadelli, P., Lanzarotti, R.: An automatic feature-based face recognition system. In: Proceedings of the 5th International Workshop on Image Analysis for Multimedia Interactive Services (WIAMIS'04) (2004)
15. Ekman, P.: Basic Emotions. Handbook of Cognition and Emotion, pp. 45–60 (1999)
16. Ortony, A., Clore, G., Collins, A.: The Cognitive Structure of Emotions. Cambridge University Press, Cambridge (1988)
17. Plutchik, R.: The nature of emotions. Am. Sci. **89**(4), 344–350 (2001)

18. Wang, W., Yang, J., Xiao, J., Li, S., Zhou, D.: Face recognition based on deep learning. In: International Conference on Human Centered Computing, pp. 812–820. International Publishing, Springer (2014)
19. Salakhutdinov, R., Hinton, G.: Deep boltzmann machines. In: Artificial Intelligence and Statistics, pp. 448–455 (2009)
20. Deng, L.: A tutorial survey of architectures, algorithms, and applications for deep learning. APSIPA Trans. Signal Inf. Process. **3**, e2 (2014)
21. Srivastava, N., Salakhutdinov, R.R.: Multimodal learning with deep boltzmann machines. In: Advances in Neural Information Processing Systems, pp. 2222–2230 (2012)
22. Căleanu, C.D.: Face expression recognition: A brief overview of the last decade. In: IEEE 8th International Symposium on Applied Computational Intelligence and Informatics (SACI), pp. 157–161 (2013)
23. Sariyanidi, E., Gunes, H., Cavallaro, A.: Automatic analysis of facial affect: A survey of registration, representation, and recognition. IEEE Trans. Pattern Anal. Mach. Intell. **37**(6), 1113–1133 (2015)
24. Danelakis, A., Theoharis, T., Pratikakis, I.: A survey on facial expression recognition in 3D video sequences. Multimedia Tools Appl. **74**(15), 5577–5615 (2015)
25. Aung, D.M., Aye, N.A.: Facial expression classification using histogram based method. In: International Conference on Signal Processing Systems (2012)
26. Murthy, G.R.S., Jadon, R.S. Recognizing facial expressions using eigenspaces. In: IEEE International Conference on Computational Intelligence and Multimedia Applications. **3**, pp. 201–207 (2007)
27. Thai, L.H., Nguyen, N.D.T., Hai, T.S.: A facial expression classification system integrating canny, principal component analysis and artificial neural network.(2011) *arXiv preprint* arXiv:1111.4052
28. Perikos, I., Ziakopoulos, E., Hatzilygeroudis, I.: Recognizing emotions from facial expressions using neural network. In: IFIP International Conference on Artificial Intelligence Applications and Innovations, pp. 236–245. Springer, Heidelberg (2014)
29. Perikos, I., Ziakopoulos, E., & Hatzilygeroudis, I.: Recognize emotions from facial expressions using a SVM and neural network schema. In: Engineering Applications of Neural Networks, pp. 265–274. Springer International Publishing, (2015)
30. Anderson, K., McOwan, P.W.: A real-time automated system for the recognition of human facial expressions. IEEE Trans. Syst. Man Cybern. Part B (Cybern.), **36**(1), 96–105 (2006)
31. Přinosil, J., Smékal, Z., Esposito, A.: Combining features for recognizing emotional facial expressions in static images. In: Esposito, A., Bourbakis, N.G., Avouris, N., Hatzilygeroudis, I. (eds.) Verbal and Nonverbal Features of Human-Human and Human-Machine Interaction, pp. 56−69. Springer, Heidelberg (2008)
32. Shinohara, Y., Otsu, N.: Facial expression recognition using fisher weight maps. In: Proceedings Sixth IEEE International Conference on Automatic Face and Gesture Recognition, IEEE. pp. 499–504 (2004)
33. Yang, J., Zhang, D., Frangi, A.F., Yang, J.Y.: Two-dimensional PCA: a new approach to appearance-based face representation and recognition. IEEE Trans. Pattern Anal. Mach. Intell. **26**(1), 131–137 (2004)
34. Oh, S.K., Yoo, S.H., Pedrycz, W.: Design of face recognition algorithm using PCA-LDA combined for hybrid data pre-processing and polynomial-based RBF neural networks: Design and its application. Expert Syst. Appl. **40**(5), 1451–1466 (2013)
35. Mohammadi, M.R., Fatemizadeh, E., Mahoor, M.H.: PCA-based dictionary building for accurate facial expression recognition via sparse representation. J. Vis. Commun. Image Represent. **25**(5), 1082–1092 (2014)
36. Tang, Y.: Deep learning using linear support vector machines.(2013). *arXiv preprint* arXiv:1306.0239
37. Ionescu, R.T., Popescu, M., Grozea, C.: Local learning to improve bag of visual words model for facial expression recognition. In: Workshop on Challenges in Representation Learning, ICML (2013)

38. Mollahosseini, A., Chan, D., & Mahoor, M.H. (2016, March). Going deeper in facial expression recognition using deep neural networks. In: IEEE Winter Conference on Applications of Computer Vision (WACV), IEEE. pp. 1–10 (2016)
39. Goodfellow, I.J., Erhan, D., Carrier, P.L., Courville, A., Mirza, M., Hamner, B., Zhou, Y.: Challenges in representation learning: A report on three machine learning contests. In: International Conference on Neural Information Processing pp. 117–124. Springer, Heidelberg (2013)
40. Szegedy, C., Liu, W., Jia, Y., Sermanet, P., Reed, S., Anguelov, D., Erhan, D., Vanhoucke, V., Rabinovich, A.: Going deeper with convolutions. (2014) *arXiv preprint* arXiv:1409.4842
41. Dean, J., Corrado, G., Monga, R., Chen, K., Devin, M., Mao, M., Ng, A.Y.: Large scale distributed deep networks. In: Advances in Neural Information Processing Systems, pp. 1223 –1231 (2012)
42. Krizhevsky, A., Sutskever, I., Hinton. G.E.: ImageNet classification with deep convolutional neural networks. Part of: Adv. Neural Inf. Process. Syst. NIPS, **25** (2012)
43. Jia, Y., Shelhamer, E., Donahue, J., Karayev, S., Long, J., Girshick, R., Darrell, T.: Caffe: Convolutional architecture for fast feature embedding. In: Proceedings of the 22nd ACM International Conference on Multimedia, pp. 675–678. ACM (2014)
44. By Aphex34 (Own work) [CC BY-SA 4.0 (http://creativecommons.org/licenses/by-sa/4.0)], via Wikimedia Commons
45. I. J. Goodfellow, D. Erhan, P. L. Carrier, A. Courville, M. Mirza, B. Hamner, W. Cukierski, Y. Tang, D. Thaler, D.-H. Lee, Y. Zhou, C. Ramaiah, F. Feng, R. Li, X. Wang, D. Athanasakis, J. Shawe-Taylor, M. Milakov, J. Park, R. Ionescu, M. Popescu, C. Grozea, J. Bergstra, J. Xie, L. Romaszko, B. Xu, Z. Chuang, and Y. Bengio, Challenges in representation learning: A report on three machine learning contests, Neural Networks, vol. 64, pp. 59–63 (2015)

Analysis of Biologically Inspired Swarm Communication Models

Musad Haque, Electa Baker, Christopher Ren, Douglas Kirkpatrick and Julie A. Adams

Abstract The biological swarm literature presents communication models that attempt to capture the nature of interactions among the swarm's individuals. The reported research derived algorithms based on the metric, topological, and visual biological swarm communication models. The evaluated hypothesis is that the choice of a biologically inspired communication model can affect the swarm's performance for a given task. The communication models were evaluated in the context of two swarm robotics tasks: search for a goal and avoid an adversary. The general findings demonstrate that the swarm agents had the best overall performance when using the visual model for the search for a goal task and performed the best for the avoid an adversary task when using the topological model. Further analysis of the performance metrics by the various experimental parameters provided insights into specific situations in which the models will be the most or least beneficial. The importance of the reported research is that the task performance of a swarm can be amplified through the deliberate selection of a communications model.

Keywords Artificial swarms · Robotics tasks

M. Haque (✉) · E. Baker · C. Ren · D. Kirkpatrick · J.A. Adams
Department of Electrical Engineering and Computer Science, Vanderbilt University, Nashville, TN, USA
e-mail: musad.a.haque@vanderbilt.edu

E. Baker
e-mail: electa.a.baker@vanderbilt.edu

C. Ren
e-mail: christopher.ren@vanderbilt.edu

D. Kirkpatrick
e-mail: kirkpa48@msu.edu

J.A. Adams
e-mail: julie.a.adams@oregonstate.edu

I. Hatzilygeroudis and V. Palade (eds.), *Advances in Hybridization of Intelligent Methods*, Smart Innovation, Systems and Technologies 85,
https://doi.org/10.1007/978-3-319-66790-4_2

1 Introduction

Animals that live in groups gain reproductive advantages, benefit from reduced predation risks, and forage efficiently through group hunting and the distribution of information amongst group members [20]. The collective behavior of these biological systems, for instance, trail-forming ants, schooling fish, and flocking birds, display tight coordination that appears to emerge from local interactions, rather than through access to global information or a central controller [8]. Numerical simulations based solely on local interaction rules can recreate coordinated movements of biological systems living in groups [2, 12, 17, 19, 27, 32].

Proposed communication models for group behavior in animals include the metric [11], the topological [1, 3], and the visual models [31]. The metric model is directly based on spatial proximity: two individuals interact if they are within a certain distance of one another [11]. Ballerini et al.'s [3] topological model requires each individual to interact with a finite number of nearest group members. The visual model, which is based on the sensory capabilities of animals, permits an individual to interact with other agents in its visual field [31]. The communication model is an important element in collective behavior, because it reveals how information is transferred in the group [31].

The development of communication networks is described as "one of the main challenges" in swarm robotics [18]. Bio-inspired artificial swarms inherit desirable properties from their counterparts in nature, such as decentralized control laws, scalability, and robustness [6]. Robustness in the context of this paper implies that the failure of one agent does not lead to the failure of the entire swarm. Despite the beneficial properties, a poorly designed communication network to an artificial swarm can lead to undesirable consequences, such as the swarm fragmenting into multiple components [18].

The evaluated hypothesis is that the three communication models—metric, topological, or visual—when used by a tasked artificial swarm will affect the swarm's performance. The evaluation analyzes how the communication models impact swarm performance for two swarm robotics tasks: searching for a goal and avoiding an adversary. The findings demonstrate that there is a significant impact of the communication model on task performance, which implies that the task performance of a deployed artificial swarm is amplified through performance-based selection of communication models.

Section 2 provides related work. Section 3 describes the coordination algorithms derived from the biological models. Experiments are presented in Sects. 4 and 5. Practical applications are discussed in Sect. 6, and an overall discussion with concluding remarks is provided in Sect. 7.

2 Related Work

Comparative evaluations of swarm communication models can be grouped into three fields: biology [3, 31], physics [4, 29], and computer science [16].

Prior research compared the communication models to identify which model best explains the propagation of information within biological species. Stranburg-Peshkin et al. [31] reported that for golden shiners, *Notemigonus crysoleucas*, the visual model best predicts information transfer within the school. The Metric and topological models were compared for flocks of European starlings, *Sturnus vulgaris* [3], and the topological model most accurately described the starlings' information network. The experiment compared the cohesion of simulated swarms using the topological and metric models, and the topological model generated more cohesive swarms [3].

Physics-based approaches compared the metric and topological models and presented the resulting system properties. Specifically, Shang and Bouffanais [29] presented results on the probability of reaching a consensus. Barberis and Albano [4] analyzed the difference in group orders (alignment and moment) that arise when using the metric and topological models.

Computer science results include evaluating the metric and topological models in the context of human-swarm interaction [16]. The human steered the swarm by manipulating a leader agent that directly influenced other swarm members. It was determined that a human can more easily manage a swarm using the topological model.

The presented evaluation appears to be the first to compare the metric, topological, and visual models for tasks on artificial agents.

3 Coordination Algorithms

The agents are modeled as *2D* self-propelled particles. A self-propelled particle is controlled through updates to its velocity heading, which in turn affects the particle's position [12, 15, 32].

The artificial agents are indexed 1 through N, where N is the number of agents in the swarm. If there is a communication link from agent $i \in \{1, \ldots, N\}$ to agent $j \in \{1, \ldots, N\}$, where $i \neq j$, agent j is a neighbor of agent i. The neighbor set of agent i, denoted by $\mathcal{N}_i(t)$ is the collection of all the neighbors of agent i at time t.

The coordination of the swarm agents is designed through a multi-level coordination algorithm. At the higher level of abstraction, an agent's neighbors are determined by the communication model. Thus, for agent i, the communication model constructs the set $\mathcal{N}_i(t)$ at each time t. At the lower abstraction level, agents only interact with their neighbors and the nature of this interaction is governed by three rules: repulsion, orientation, and attraction. The rules are based on Reynolds's rules for boids (see [27]), which are similar to the biological swarm literature (e.g., [2]).

Each agent's zones of repulsion, orientation, and attraction are centered at the agent's position and are parameterized through the radii r_{rep}, r_{ori}, and r_{att}, respectively, where $r_{rep} < r_{ori} < r_{att}$. The zones are represented as circles in the 2D case.

The heading of each agent $i \in \{1, \ldots, N\}$ is updated as follows: (1) Veer away from all agents in $\mathcal{N}_i(t)$ within a distance r_{rep}, (2) Align velocity with all agents in $\mathcal{N}_i(t)$ that are between a distance of r_{rep} and r_{ori}, and (3) Remain close to all agents $j \in \mathcal{N}_i(t)$ that are between a distance of r_{ori} and r_{att} [21, 27].

3.1 Communication Models

The metric model uses a single parameter d_{met} that represents a distance measure. All agents within a distance d_{met} from agent i are i's neighbors, as shown in Fig. 1a. Due to the symmetric nature of the model, if $j \in \mathcal{N}_i(t)$, then $i \in \mathcal{N}_j(t)$. A stochastic version of this model was developed to analyze starling data [5]. The analyzed models assign neighbors in a deterministic manner.

The topological model is characterized by n_{top}, measured in units of agents. $\mathcal{N}_i(t)$ is the set containing the n_{top} nearest agents from agent $i \in \{1, \ldots, N\}$. Zebrafish, *Danio rerio*, have 3–5 topological neighbors [1], and starlings coordinate, on average, with the nearest 6–7 birds [3]. Figure 1b depicts the neighbors of agent i, with n_{top} set to 5.

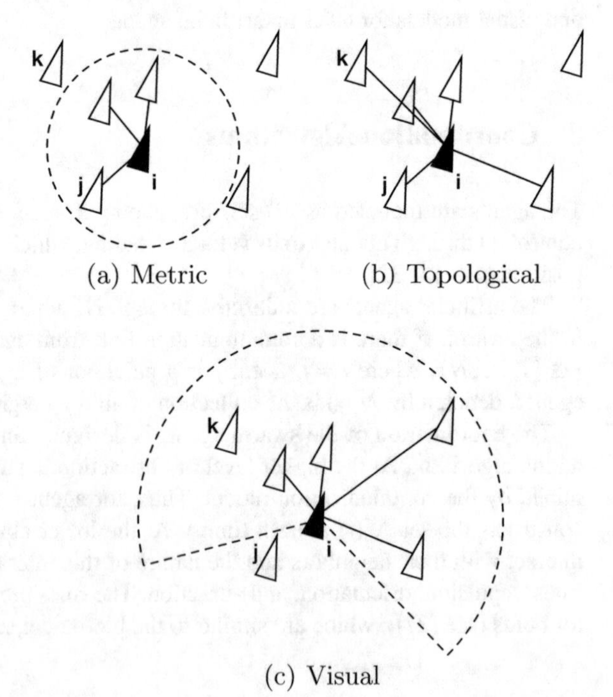

Fig. 1 Agents (*triangles*) are shown in relation to the focus agent (*filled triangle*), labeled i. The communication links from agent i to its neighbors are represented with lines. **a** Metric: Agent k is at a distance greater than d_{met} (*dashed circle*) from agent i; **b** Topological: n_{top} is set to 5; **c** Visual: The visual range of agent i is shown (*dashed sector*), where agent j is in agent i's blindspot and agent k is occluded from agent i by another agent

A sensing range, a blindspot, and occlusion are used to describe the visual model [31]. Agent j is a neighbor of agent i, if three conditions are met: (1) The distance between the two agents is less than d_{vis}, (2) Agent j is not in agent i's blindspot, and (3) The line-of-sight between the agents is not occluded by another agent or object in the environment. A blindspot emerges because the agent's sensing range is characterized by an angle $\pm\phi$ from its heading [12, 15]. Figure 1c depicts agent i's sensing range, with ϕ set to $2\pi/3$ radians.

The particular choices made for the values of d_{met}, n_{top}, d_{vis}, and ϕ can be characterized as inheriting from the "descriptive agenda" of multi-agent learning [22, 30]. The goal in the descriptive agenda is to model the underlying phenomenon from the social sciences (biological swarm communication models). The biological swarm literature provides parameter values that are used to compare the different communication models on tasked artificial swarms. d_{met} was set to r_{att} for metric model experiments (e.g., [2, 11]). The visual model experiments set an agent's d_{vis} to half the size of the diagonal of the world with $\phi = 2\pi/3$ radians [12, 31]. $n_{top} \in \{5, 6, 7, 8\}$ for the topological experiments, allowing some variability, while remaining close to what was observed in nature [3].

The novelty is the comparative evaluations of the different communication models that are *solely* based on the biological swarm literature; hence, strictly inheriting from a descriptive agenda. Traditional artificial swarm communication models do not typically mimic the three communication models (e.g., [9]). Although, perception-based models that rely on line-of-sight communication, such as a swarm of foot-bots responding to light sensors, is a variant of the visual model [14]. As such, one potential application is to serve as a guide for hardware selection.

4 The Search for a Goal Experiment

4.1 Experimental Design

All experiments were conducted using the `Processing` open-source programming language on a 8 GB, 2.6 GHz Intel Core i5 Macbook Pro. The body length, BL, of each agent was set to 2 pixels. The size of the world was 600×600 pixels.

The communication model is the primary independent variable: metric, topological, and visual. Additional independent variables were: the number of agents, the number of obstacles, the radius of repulsion, the radius of orientation, and the radius of attraction. The experiment combined each of the primary independent variables with each of the additional independent variables. The resulting pair-wise combinations offers a more comprehensive analysis of the effect of the communication models.

The number of *agents*, N, was 50, 100, and 200. The tuple $(r_{rep}, r_{ori}, r_{att})$ describes an agent's repulsion, orientation, and attraction zones. The radius of repulsion, r_{rep}, was set to either $5 \times BL$ or $10 \times BL$. The radius of orientation, r_{ori}, was assigned to

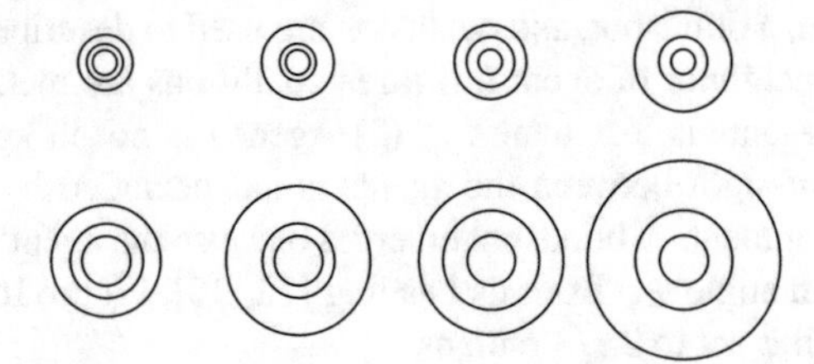

Fig. 2 The eight possible interaction zone configurations. The inner-most, middle, and outer-most *circles* represent the zones of repulsion, orientation, and attraction, respectively

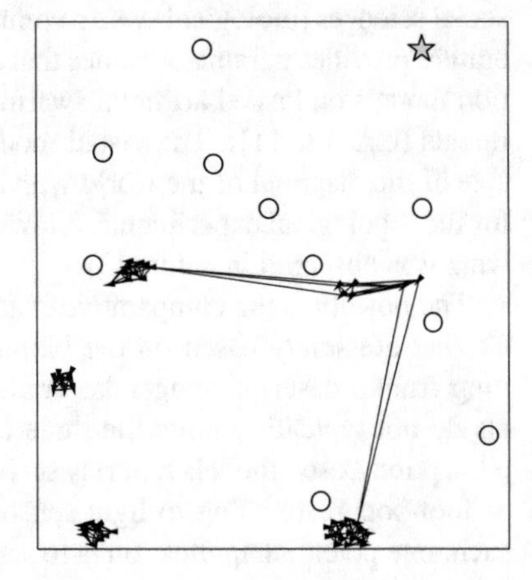

Fig. 3 An artificial swarm performing the search for a goal task using the topological model. The center of the goal area is represented by a *star*, *circles* represent obstacles, agents are *filled triangles*, and the *lines* denote communication links. The trial parameters were: $N = 50$, $N_{obs} = 0.20N$, $r_{rep} = 20$, $r_{ori} = 40$, $r_{att} = 60$, and $n_{top} = 6$

either $1.50 \times r_{rep}$ or $2.00 \times r_{rep}$, and the radius of attraction, r_{att}, was given a value of either $1.50 \times r_{ori}$ or $2.00 \times r_{ori}$. Designing the interaction zones in this manner results in 2^3 possible tuples with varying (relative) zone sizes, as illustrated in Fig. 2.

The search for a goal task included *environmental* obstacles. The number of obstacles, N_{obs}, was 0%, 10%, or 20% of N.

The objective of the artificial swarm during the *search for a goal* is to locate a single goal location,[1] the star in Fig. 3. The goal area's size is scaled to ensure the swarm is able to fit within the goal area. The world is bounded by a wall that exerts a repulsive force. An agent can sense the goal if it is within r_{att} of the goal area's location. Once an agent locates the goal, it can communicate the location to its neighbors. Agents aware of the goal's location update their headings by equally weighing the desire to travel to the goal and the desire to follow the interaction rules, which was employed by Couzin et al. [11] and Goodrich et al. [16]. The simulation runs for 1,000 iterations.

The *percent reached*, denoted by R, determines the number of agents that reached the goal area, expressed as a percentage of the swarm's size, N, at the end of the task.

[1] Videos of example trials can be found at http://www.eecs.vanderbilt.edu/research/hmtl/wp/index.php/research-projects/human-swarm-interaction/emulating-swarm-communications/.

The *latency*, L, measures the rate of information transfer in the swarm during the task. Specifically, latency represents the number of iterations required for the swarm to transition from a state where at least one agent knows the goal's location to all agents being aware of the goal's location. Degenerate cases are processed by setting the latency to the maximum possible duration, 1,000 iterations, Based on this definition, the simulator did not influence this metric.

The clustering coefficient is the fraction of pairs of a swarm agent's neighbors that are neighbors with each other [13]. The coefficient ranges from 0, where none of the swarm agent's neighbors are neighbors with each other, to 1, where all pairs of a swarm agent's neighbors are neighbors with each other. The *swarm clustering coefficient*, denoted by SCC, averages the clustering coefficients of all swarm agents. A high swarm clustering coefficient implies a dense communication network and redundant information passing between the agents. While calculating the swarm clustering coefficient, the asymmetric nature of the communication links that resulted from the topological and visual models were ignored. Strandburg-Peshkin et al. [31] performed the same treatment on directed links when comparing this metric across different communication models for fish data. This metric permits comparison to prior findings.

The three hypotheses for this task are:

1. H_{sg1}: $R_V > R_T > R_M$,
2. H_{sg2}: $L_V < L_T < L_M$, and,
3. H_{sg3}: $SCC_V < SCC_T < SCC_M$.

The subscripts associated with the performance metrics indicate the metric (M), the topological (T) and the visual (V) models.

Hypothesis H_{sg1} assumes that a greater percentage of agents will reach a goal using the visual model and that the metric model will have the lowest percentage reached. The hypothesis is based on the *potentially* long-range sensing capabilities associated with the visual model. Agents favorably oriented and not occluded by obstacles or other agents have a higher chance of communicating with an agent that has located the goal. Moreover, fewer stragglers may arise with the visual and topological models, thus increasing the percent reached. H_{sg1} further assumes that the limit on n_{top}, compared to the range of d_{vis}, allows a greater percentage of agents to arrive at a goal using the visual model, compared to the topological model.

Establishing long-range communication between two agents in the visual model depends on the orientation of the agents and occluding factors. The range d_{vis} may not be a limiting factor in identifying neighbors when positioned in the interior of the swarm. However, any occurrence, regardless of how infrequent, of a long-range link in the network can act as a short-cut for transferring information. As such, H_{sg2} states that information diffuses faster in swarms using the visual and topological models, than with the metric model.

Hypothesis H_{sg3} states that the swarm clustering coefficient will be the highest in the metric model and the lowest in the visual model. Communication links in the metric and topological models are not affected by occlusions, a factor that is expected to yield sparser networks for the visual model.

A trial is defined as a single simulation run for a given selection of parameters, $(N, N_{obs}, r_{rep}, r_{ori}, r_{att})$. Twenty-five trials for each parameter selection were completed. The total number of trials for the search for a goal task was 10,800: 1,800 trials for each of the metric and visual models, and 7,200 trials for the topological model (1,800 trials for each of the four values of n_{top}).

4.2 Results

The Anderson-Darling test for normality indicated that all performance metrics: percent reached ($A = 431.01, p < 0.001$), latency ($A = 621.88, p < 0.001$), and swarm clustering coefficient ($A = 162.72, p < 0.001$) were distributed normally. An analysis of variance (ANOVA) by n_{top} did not find a significant difference for the topological model's performance. Without loss of generality, the topological trials with $n_{top} = 7$ are used in the reported ANOVAs.

The topological and visual models had virtually identical mean **percent reached**, as reported in Table 1. The ANOVA found that model type had a significant impact on the percent reached ($F(2, 5398) = 83.91, p < 0.001$). A Fisher's LSD test investigated the pair-wise differences. There was no significant difference between the visual and topological models, and the metric model had a significantly lower percent reached compared to the other models.

All data was further analyzed by the number of agents, number of obstacles, and the radii of repulsion, orientation, and attraction. ANOVAs showed significant interactions between the communication models and the number of agents ($F(2, 5398) = 11.26, p < 0.001$), the number of obstacles ($F(2, 5398) = 8.85, p < 0.001$), and the radius of attraction ($F(2, 5398) = 2.52, p = 0.043$). No significant interactions were found for the radii of orientation and repulsion.

Table 1 The search for a goal task descriptive statistics by models. The best means are in bold. (The percent reached, latency, and swarm clustering coefficient are represented by *R*, *L*, and *SCC*, respectively)

Model	Statistic	R	L	SCC
Metric	Mean	27.68	637.79	0.95
	Median	0.00	1000.00	0.95
	Std. Dev.	41.60	471.73	0.03
Topological	Mean	**39.08**	864.99	0.62
	Median	34.00	1000.00	0.62
	Std. Dev.	31.75	290.20	0.06
Visual	Mean	**41.10**	**438.73**	**0.31**
	Median	22.00	31.00	0.33
	Std. Dev.	42.56	487.99	0.07

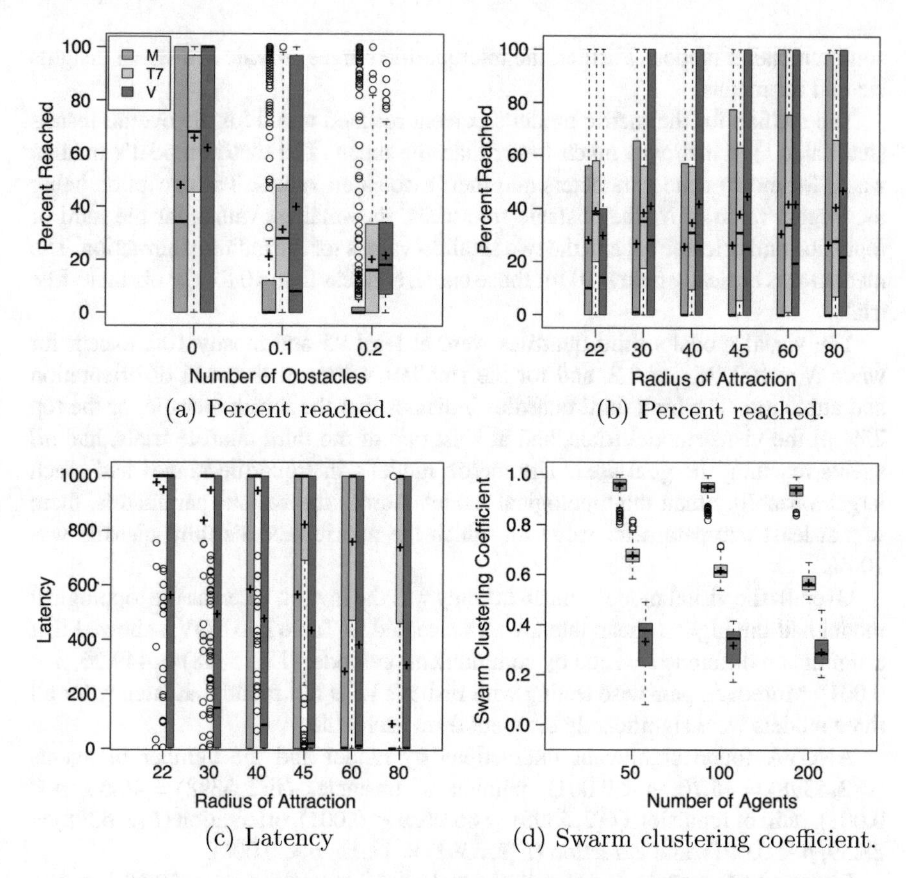

(a) Percent reached. (b) Percent reached.

(c) Latency (d) Swarm clustering coefficient.

Fig. 4 The search for a goal task performance metrics. Each box plot denotes the first and third quartile of data. The *horizontal lines* indicate the medians, the crosses represent the means, and the *circles* show the outlying data. The legend for the plots **b–d** can be found in **a**, where *M*, *T*7, and *V* denote the metric, topological (with $n_{top} = 7$), and visual models, respectively

Fisher's LSD test showed that for $N = 50$, there was no significant difference between the visual and topological models. The mean percent reached was the highest when $N = 100$ using the topological model. The visual model had the highest mean percent reached for $N = 200$. The percent reached for the metric model was significantly different compared to the other models across all values of N.

The mean percent reached for all the models decreased as additional obstacles were included, as shown in Fig. 4a.

At $r_{att} = 22.50$ there was no significant difference in percent reached for the metric and visual models, see Fig. 4b. The metric model's mean percent reached was significantly higher at $r_{att} = 22.50$ compared to $r_{att} = 80$.

The means are susceptible to the influence of outliers, thus the median values are also reported as a central tendency measure to better assess the performance of the communication models. Further, the interquartile ranges provide additional insights beyond the means.

The median for the metric model's percent reached was 0 for the overall results (see Table 1), which was much lower than the mean. The metric model's median was 0 for most of the parameters and their associated values. The exception being the largest value of N, the obstacle-free trials, the smallest values for the radii of repulsion and orientation, and the two smallest values for the radius of attraction. The median was typically below 10 for those cases, and less than 40 for the obstacle-free trials.

The visual model's third quartiles were at least 95 and mostly 100, except for when $N = 100$, $N_{obs} = 0.2$, and for the smallest values of the radii of orientation and attraction. The high third quartiles indicates that the fourth quartile, or the top 25% of the visual model trials, and at least one of the third quartile trials, had *all* agents reaching the goal area. The metric model's interquartile ranges had much larger variability than the topological model. Across the various parameters, there was at least one parameter value for which the metric model's third quartile was 100%.

Overall, the visual model's mean **latency** was the lowest, whereas the topological model had the highest mean latency, as presented in Table 1. ANOVA showed that a significant difference existed by communication model ($F(2, 5398) = 449.26, p < 0.001$). Moreover, pair-wise testing with Fisher's LSD test found that latency for all three models were significantly different from each other.

ANOVA found significant interactions by model and the number of agents ($F(2, 5398) = 45.70$, $p < 0.001$), number of obstacles ($F(2, 5398) = 40.60$, $p < 0.001$), radii of repulsion ($F(2, 5398) = 66.96, p < 0.001$), orientation ($F(2, 5398) = 28.59, p < 0.001$), and attraction ($F(2, 5398) = 11.15, p < 0.001$).

Fisher's LSD test showed that the visual model latency at $r_{att} = 22.50$ was significantly lower than the metric and topological models. At $r_{att} = 80$, the analysis found a significant difference across each of the models, with the metric model's mean latency being lowest (see Fig. 4c). An identical trend occurs for the lowest and highest radius of orientation.

The metric model's median latency was 1000, for most cases across the number of agents, number of obstacles, and the radii of repulsion, orientation, and attraction. The exceptions occurred for the largest value of the radius of repulsion, the two largest values of the radius of orientation, and the two largest values of the radius of attraction, as shown in Fig. 4c. The median latency was typically 0 for those exceptional cases. Similarly, the topological model's median latency was 1000 across the variables. Additionally, the first quartile of the topological model's latency was 1000 in most cases, and in certain cases, it was at least greater than 400 (see Fig. 4c). The visual models' median latency was lower than the mean, and was 0 for the largest value of the number of agents, the largest value of the radius of repulsion, and the two largest values of the radii of orientation and attraction (see Fig. 4c).

The mean **swarm clustering coefficient** was lowest in the visual model and highest in the metric model. An ANOVA showed a significant difference by model

($F(2,5398) = 1810, p < 0.001$). Fisher's LSD test found that all the models had significantly different means.

Results from ANOVA showed that for the swarm clustering coefficient, there were significant interactions by model and the number of agents ($F(2,5398) = 631.50$, $p < 0.001$), the number of obstacles ($F(2,5398) = 2132.00, p < 0.001$), the radii of repulsion ($F(2,5398) = 320.90$, $p < 0.001$), orientation ($F(2,5398) = 144.40$, $p = 0.03$), and attraction ($F(2,5398) = 166.40, p < 0.001$). The results of Fisher's LSD test found a significant pair-wise difference between the models across all variables and associated values.

The median swarm clustering coefficients for all communication models were generally close to the means across all parameters and associated values. The interquartile ranges were typically tight, with only a few cases where the maximum value of one model overlapped with the minimum value of another. Those cases were the smallest number of agents (see Fig. 4d), the smallest radii of repulsion, orientation, and attraction.

4.3 Discussion

H_{sg1} was partially supported. The topological and visual models outperformed the metric model in reaching the goal area, yet there was no clear difference between the visual and topological models.

The visual model latency was substantially lower than the topological and metric models; however, the metric model outperformed the topological model in terms of the transfer of information. As such, H_{sg2} was also only partially supported. The metric model's bidirectional communication links possibly allowed information to spread faster through the network, compared to the topological model.

Similar to Strandburg-Peshkin et al.'s [31] results for fish, the swarm clustering coefficient was lowest with the visual model. The clustering coefficient for fish with the topological model was higher than the metric model, contrary to the findings presented in Table 1. One possible reason for this difference can be attributed to the difference in using collective motion experimental data as opposed to modeling through self-propelled particles.

Based on the general findings, the visual communication model is the best for artificial swarms completing a search for a goal task when fewer redundant connections are desired, because it resulted in virtually the best percent reached, the lowest latency, and the lowest swarm clustering coefficient. A low swarm clustering coefficient can be disadvantageous in noisy environments, which can benefit from redundant communication links. The metric and topological models are preferred for such environments, because of their high swarm clustering coefficients. Furthermore, given a noisy environment and a requirement for only a few agents to reach the goal, then the metric model is preferred. Given the same noisy environment, but a high percentage of agents needed to reach the goal, then the topological model can be used. The tradeoff is the model's high latency.

The analysis by the radius of attraction, which was the value of d_{met}, revealed that the metric and visual models are fundamentally different from one another and the difference does not stem from the visual model's larger communication range. Overall, the visual-based swarms performed better than the metric-based swarms. However, at the lowest value of the radius of attraction ($d_{met} = 22.50$), the metric and visual models had comparable mean percent reached. Furthermore, for the highest value of the radius of attraction, or $d_{met} = 80$, the latency of the metric model was shown to be significantly lower than the visual, which used a range of $d_{vis} = 425$.

5 The Avoid an Adversary Experiment

5.1 Experimental Design

This experiment was performed using the same machine and the experimental parameters, other than N_{obs}, were identical. No obstacles were included in this experiment.

The swarm is required to avoid a predator-like agent[2] during the *avoid an adversary* task, which is modeled through a repulsive force exerted by the adversary on the swarm agents [3]. The swarm (dark mass in Fig. 5a) is initially aligned facing the predator (triangle in Fig. 5a). The predator is the same size as the swarm agents and can occlude the visual communication between agents. For illustrative purposes, the rendering of the predator has been increased. The predator (moving in a predefined path) and swarm travel toward each other and when the swarm agents are within r_{att} of the adversary, the predator's repulsive forces affect the swarm agents' heading. Agent positions are initially distributed in an area that is proportional to the swarm's size, N. The predator's starting position is horizontally offset, such that the predator and swarm travel the same distance to meet, regardless of the swarm's size. The effects of the adversary on the swarm are isolated by removing the environmental obstacles and negating the wall's repulsive forces. Each trial runs for 200 iterations.

Dispersion, denoted by D, is measured as the percentage increase of the average agent to agent distance from the start to the end of the trial. The average agent to agent distance has significance in the biological literature and is one of eleven parameters considered when characterizing the emergent properties of fish [26].

A connected component is defined as the largest collection of agents in which any two agents are either connected directly by a communication link or indirectly via neighbors [13]. The *number of connected components*, CCO, is calculated at the end of a trial, and is 1 at the start of a trial.

The *percent isolated components*, represented by I, is the percentage of swarm agents that have no neighbors.

[2] Videos of example trials can be found at http://eecs.vanderbilt.edu/research/hmtl/wp/index.php/research-projects/human-swarm-interaction/emulating-swarm-communications/.

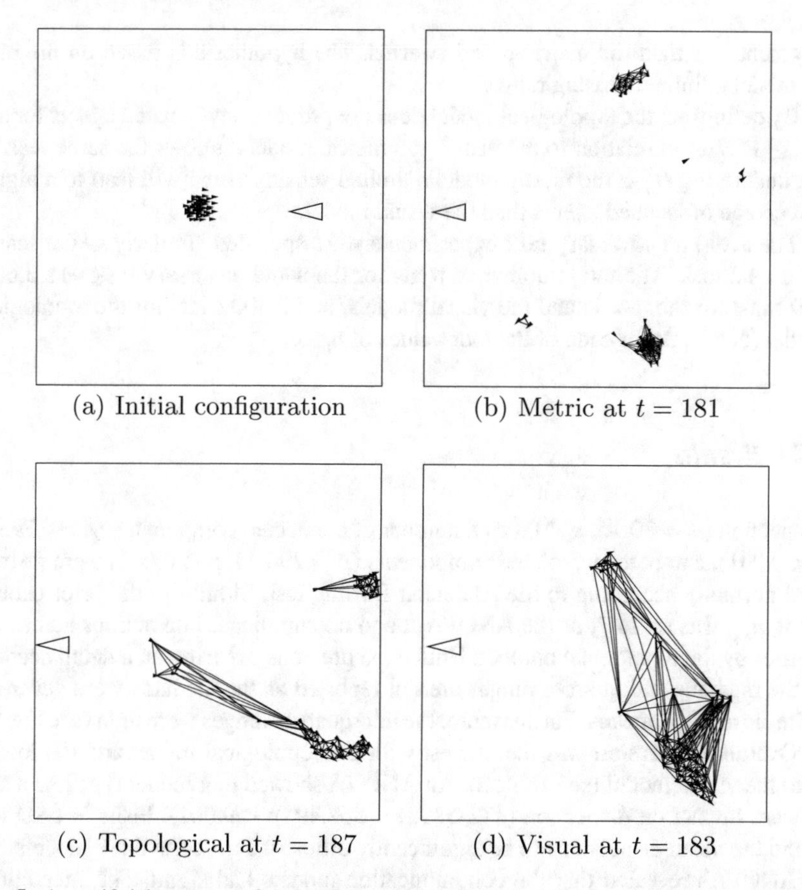

Fig. 5 An artificial swarm performing the avoid an adversary task under all three communication models. The adversary is denoted by a *triangle* and swarm's agents are represented by *filled triangles*. The *lines* between agents denote communication links. The trial parameters were: **b–d** $N = 50$, $r_{rep} = 10$, $r_{ori} = 15$, $r_{att} = 30$; **b** $d_{met} = 30$; **c** $n_{top} = 6$; **d** $d_{vis} = 425$

The three hypotheses for this task are:

1. H_{aa1}: $D_V < D_T < D_M$.
2. H_{aa2}: $CCO_V < CCO_T < CCO_M$.
3. H_{aa3}: $I_T < I_V < I_M$.

The subscripts indicate the communication models.

Hypothesis H_{aa1} assumes that the metric model will generate swarms with the highest dispersion due to fragmentation. Additionally, the topological and visual models are expected to attract outlying agents back into the main swarm after the adversary's attack, reducing the swarm's dispersion.

Hypothesis H_{aa2} states that swarms using the visual model will fragment into fewer connected components compared to the topological swarms, which will fragment less than the metric-based swarms. The hypothesis is based on the metric model's limited sensing range.

By definition, the topological model does not produce any isolated agents for any $n_{top} \geq 1$. H_{aa3} in relation to the visual and metric models follows the same reasoning underlying H_{aa2}: the metric model's limited sensing range will lead to a higher percentage of isolated agents than the visual model.

The avoid an adversary task experiments were specified similarly to the search for a goal task. The total number of trials for the avoid adversary task was 3,600: 600 trials for the metric and the visual models, and 2,400 trials for the topological model (600 trials for each of the four values of n_{top}).

5.2 *Results*

Dispersion ($A = 70.16, p < 0.001$), number of connected components ($A = 179.90$, $p < 0.001$), and percent isolated components ($A = 296.44, p < 0.001$) were distributed normally according to the Anderson-Darling test. Similar to the prior experiment, n_{top} was set to 7, as the ANOVA found no significant interactions across the metrics by the topological number. Unlike the previous experiment, a detail account of the medians and quartile ranges are not reported as the medians were generally quite close to the means. Furthermore, the interquartile ranges were tight (see Fig. 6).

Overall, **dispersion** was the highest with the topological model and the lowest with the visual model (see Table 2). An ANOVA showed that model type had a significant impact on dispersion ($F(2, 5398) = 562.49, p < 0.001$). Fisher's LSD test found the mean dispersions to be significantly different across the three models.

ANOVAs revealed that the communication models had significant interactions for the number of agents ($F(2, 5398) = 118.32, p < 0.001$), the radius of repulsion ($F(2, 5398) = 363.27, p < 0.001$), the radius of orientation ($F(2, 5398) = 26.15, p < 0.001$), and the radius of attraction ($F(2, 5398) = 9.98, p < 0.001$).

Dispersion using the topological model was significantly higher compared to the metric and the visual models for all values of N. Fisher's LSD tests showed that the visual model dispersion was significantly lower compared to the metric model at $N = 50$. However, no significant difference between the metric and visual model dispersions was found for the other values of N (see Fig. 6a).

Fisher's LSD test found that the mean dispersion for the visual model was significantly lower than the metric model at $r_{rep} = 10$, but significantly higher than the metric model at $r_{rep} = 20$, as shown in Fig. 6b. Similarly, as the values of the radii of orientation and attraction increased, the metric model's dispersion decreased to a value significantly lower than the visual model.

ANOVA determined that model type had a significant impact on the **number of connected components** ($F(2, 5398) = 1776.23, p < 0.001$). This metric was significantly different between each of the communication models, as indicated by the Fisher's LSD test. The visual model had the lowest number of connected components, while metric had the highest, as shown in Table 2.

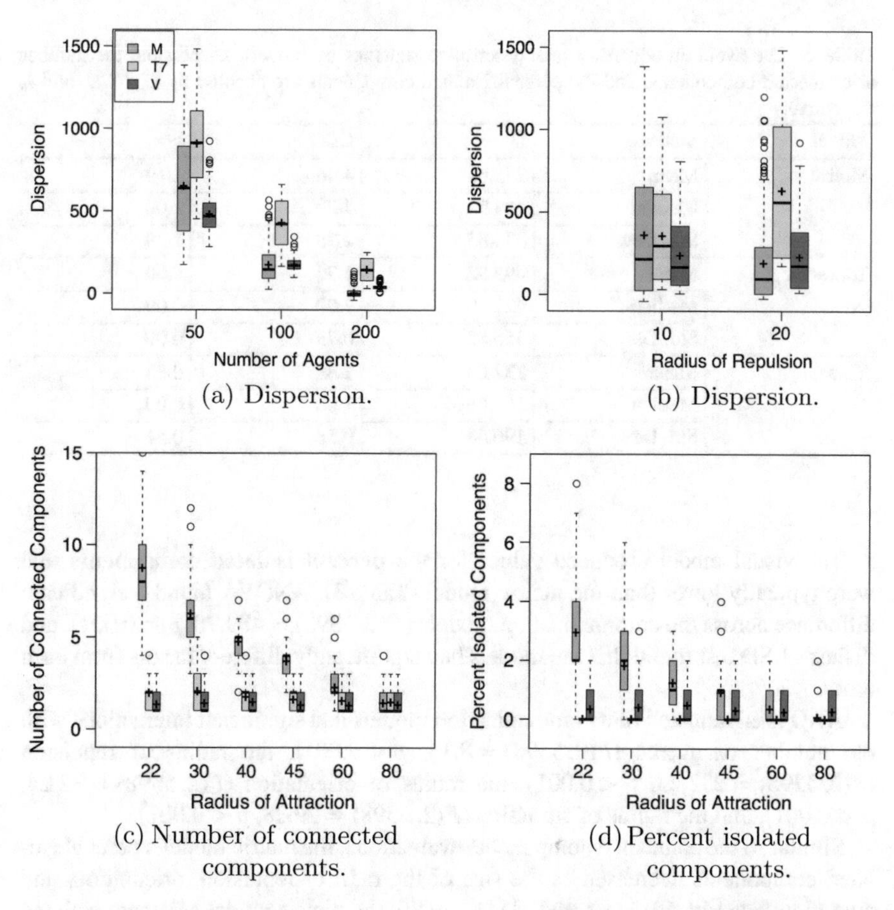

(a) Dispersion.

(b) Dispersion.

(c) Number of connected components.

(d) Percent isolated components.

Fig. 6 The avoid an adversary task performance metrics. The legend for the plots **b–d** can be found in **a**, where M, $T7$, and V denote the metric, topological (with $n_{top} = 7$), and visual models, respectively

The ANOVAs found significant interactions by the number of agents ($F(2, 5398) = 5.25$, $p < 0.01$), and the radii of repulsion ($F(2, 5398) = 772.53$, $p < 0.001$), orientation ($F(2, 5398) = 133.89$, $p < 0.001$), and attraction ($F(2, 5398) = 53.72$, $p < 0.001$).

Fisher's LSD test showed that the number of connected components was significantly different between all three models across the number of agents. Visual had the lowest number of connected components, whereas metric had the highest, for all values of N.

The metric model generated fewer connected components as the radius of attraction increased (see Fig. 6c). At $r_{att} = 80$, there was no significant difference between the metric and visual models in the number of connected components.

Table 2 The avoid an adversary task descriptive statistics by models. Dispersion, the number of connected components, and the percent isolated components are denoted by *D*, *CCO*, and *I*, respectively

Model	Statistic	D	CCO	I
Metric	Mean	275.61	4.46	1.19
	Median	144.71	4.00	1.00
	Std. Dev.	334.85	2.78	1.38
Topological	Mean	**493.92**	1.75	**0.00**
	Median	421.50	2.00	0.00
	Std. Dev.	356.32	0.79	0.00
Visual	Mean	232.03	**1.35**	0.33
	Median	168.68	1.00	0.00
	Std. Dev.	196.64	0.58	0.54

The visual model produced values for the **percent isolated components** that were typically lower than the metric model (Table 2). ANOVA found a significant difference across the communication models ($F(2, 5398) = 489.78, p < 0.001$), and Fisher's LSD test found that the models had significantly different means from each other.

ANOVAs indicated that communication models had significant interactions with the number of agents ($F(2, 5398) = 8.14,\ p < 0.001$), the radius of repulsion ($F(2, 5398) = 232.23,\ p < 0.001$), the radius of orientation ($F(2, 5398) = 32.24$, $p < 0.001$), and the radius of attraction ($F(2, 5398) = 29.28, p < 0.001$).

Similar to the connected components evaluations, the metric model's percent isolated components decreased as the size of the radii of repulsion, orientation, and attraction (see Fig. 6d) increased. At, $r_{att} = 80$, the metric model's percent isolated components was significantly lower than the visual model.

5.3 Discussion

The topological model produced the highest dispersion compared to the other models. H_{aa1} was only partially supported due to the topological's higher dispersion compared to the metric model.

H_{aa2} was fully supported, as the visual model produced the smallest number of connected components, whereas the metric model generated the highest number of connected components. The visual model's percent isolated components was lower than the metric model, which fully supports H_{aa3}.

A high dispersion in some biological species may serve to confuse a predator from singling out a particular swarm agent [3]. Thus, if a higher dispersion is preferred, the general findings indicate that the topological communication model is the best for the avoid an adversary task, because it offers the highest dispersion, paired with low connected components, and no isolated components. A high dispersion can be disadvantageous if environmental features physically constrain the swarm's movement. The metric and the visual models are preferred for such environments, as they provide a lower dispersion. However, if a task requires a low percentage of isolated components, then the visual model is preferred, otherwise, the metric communication model will suffice.

The results across independent variables did not find the visual model's relatively larger communication range to provide an unfair advantage over the metric model. At the highest radius of attraction, or $d_{met} = 80$, there was no significant difference in number of connected components between the metric and visual models, despite d_{vis} being 425.

6 Practical Applications

This research is intended to serve as the foundation of a more complete examination of what factors impact swarm performance. To date, this research has focused on the set of behavior, environment, task, and hardware (BETH) as factors likely to impact swarm performance. The behavioral components in this research were the communication model used and the values set for the radii of repulsion, orientation, and attraction. The environmental variables in this research were the number of obstacles and the number of adversaries, but future research will explore other variables such as size of the area of deployment, environmental hazards, and characteristics of adversaries. This research only considered two tasks, search for a goal and avoid adversary, but these simple tasks form the basis of many more complex tasks with both military and civilian applications [7, 10]. The number of agents deployed to a task was the primary hardware limitation in this research, but as discussed in the previous section, the radii of repulsion, orientation, and attraction can be limited by hardware capabilities; other physical factors such as agent size and speed are outside the scope of this research, but future research needs to explore the impact these factors have singularly and in concert with the other components of BETH. Identifying the BETH variables that impact swarm performance and quantifying the effects of their interactions enables the development of a decision support software to optimize the likelihood of successful task completion for any given combination of known and unknown values of the components of BETH.

The body of research that explores decision support software (DSS) extends as far back as the 1970s and encompasses many different algorithms for processing the available information [25]. The goal of such software for operators of remotely deployed mobile robots is to provide decision support by simplifying the information presented to the operator [33]. A full examination of DSS and the design of interfaces for robot operators would far exceed the available space, and the design, implementation, and validation of such an interface requires its own lengthy research process. A simplified example of design and use of such an interface is given below

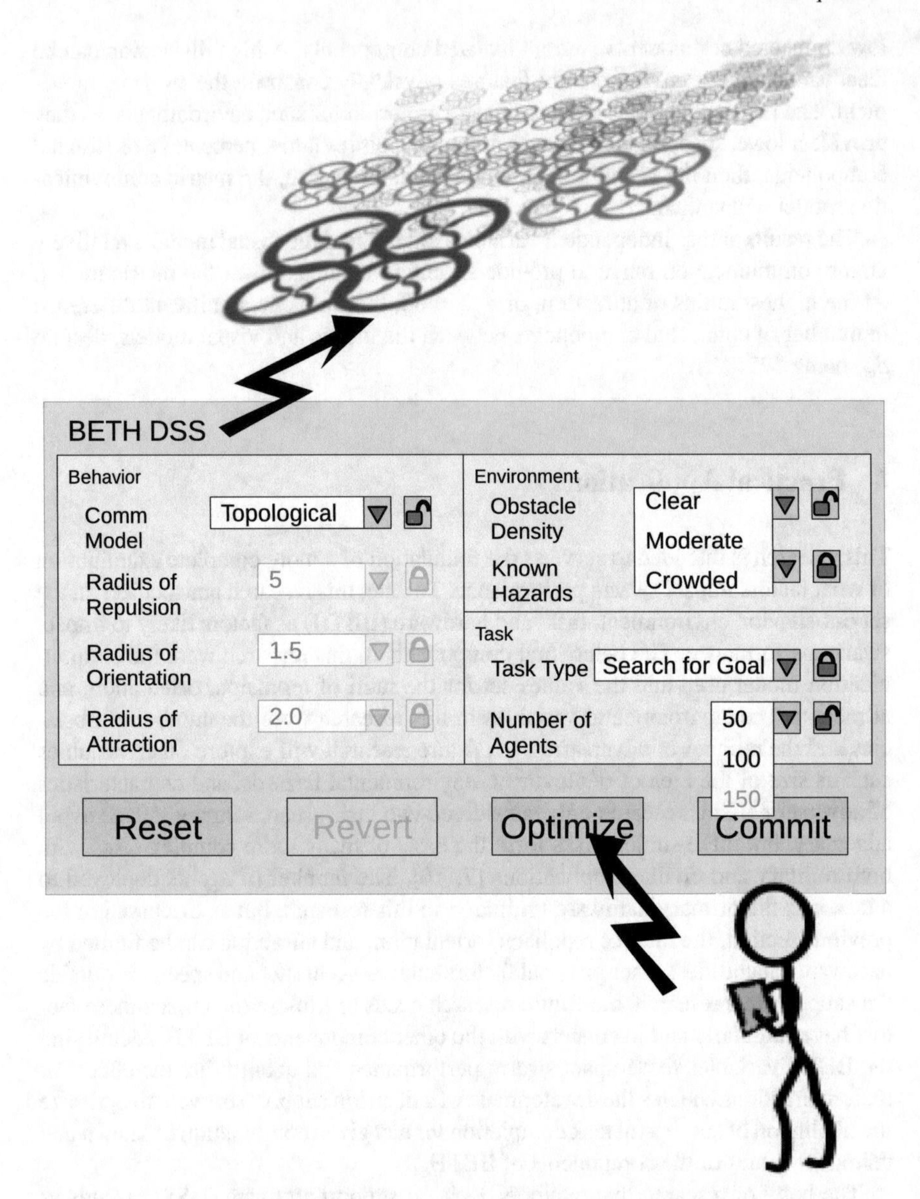

Fig. 7 An example interface using the BETH DSS model is shown. This simple wireframe illustrates how gathered data can reduce the decisions an operator needs to make in order to optimize the performance of a remotely deployed artificial swarm

to illustrate how knowledge of the factors that impact swarm performance can be used to support the operator of a robotic swarm.

Figure 7 shows an example interface that uses the BETH model to organize and simplify information for the operator deploying a swarm of robots. The available performance factors are presented grouped by Behavior, Environment, and Task. Hardware factors that impact swarm performance are made implicit by limiting the behavior, environment, and task factors to values permitted by the available hardware, freeing the operator from tracking hardware capabilities.

To use the interface, the operator must select a Task Type. The operator can then assign values for as many of the remaining performance factors as desired. For example, the operator in Fig. 7 is creating a Search for a Goal task. The operator has specified the topological communication model, but has left the padlock button toggled to "unlocked." Thus, the system the system is allowed to change the communication model during the optimization process, if a different model has a higher likelihood of success. The three radii of repulsion, orientation, and attraction are shown greyed out; the interface makes the information available but clearly indicates that the values of the radii cannot be changed by the operator (presumably the values are limited by the hardware capabilities of the swarm).

The operator has the option to supply information about the Environment where the swarm will be deployed; in this example the operator has provided a value for Known Hazards and is selecting a value for Obstacle Density. The operator has locked the Known Hazards field and left the Obstacle Density field unlocked, indicating that the system can adjust the Obstacle Density value if provided with new information, but that the information the operator has provided about he Known Hazards is to be assumed true even if the system cannot detect those hazards.

The Number of Agents can be set by the operator. The operator in the example the operator can choose between 50 and 100 agents although the value 200 is greyed out, indicating that some of the operators robots are otherwise engaged, lost, or damaged. The operator can press the Optimize button, and the system will validate the deployment variables. If the system determines that changing the values of any of the unlocked Behavior or Task variables will improve the likelihood of success, or if the system has updated information for any unlocked environmental variables, the system will update the variables with the new values, highlighting the fields that have changed so the operator understands the changes. When the operator is satisfied with the deployment configuration, pressing the "Commit" button sends the command to the swarm.

7 Discussion and Conclusion

The presented research focuses on a general hypothesis that the selection of communication model impacts the swarm's task performance. The general findings demonstrated that there was a significant impact of model type on task performance. Further, the results show that the visual model resulted in the best overall task performance for the search for a goal task, while the best overall performance was achieved with the topological model for the avoid an adversary task. The relevance of this outcome is that the intelligence of a remotely deployed swarm is amplified through the deliberate selection of a communication model. Additional analysis of typical artificial swarm tasks is necessary to fully support the general hypothesis; however, the presented results provide preliminary evidence that artificial swarm design needs to consider the communication model and task pairing in order to optimize the overall swarm performance.

Based on the presented search for a goal and avoid an adversary task results, connections can be made to the biological swarm literature. Couzin et al. [12] showed that the size of the radius of repulsion did not have an effect on the transitions between different swarm movement patterns. Rather, the relative sizes of the radius of orientation to the radius of repulsion and the radius of attraction to the radius of orientation produces the transitions. For instance, simulated swarms rotate in a torus when the ratio of the radius of orientation to the radius of repulsion is relatively low and the ratio of the radius of attraction to the radius of orientation is relatively high. Presented results for the search for a goal task conform to Couzin et al.'s [12] results in relation to the radius of repulsion. The duration of this task (1000 iterations) resulted in trials that demonstrated swarm movement patterns, as found by Couzin et al. Similar results were expected for the avoid an adversary task; however, were not found due to the task's short duration (200 iterations).

The scope of the reported research does not follow the so-called prescriptive agenda where the values of the model parameters are free design choices [22, 30]; thus, d_{vis} is not varied. This line of inquiry will become necessary when prescribing the communication models to specific platforms, such as the s-bots, which are equipped with proximity and vision sensors [24]. Analyzing the effects of varying model parameters, such as d_{met} and d_{vis} will also be necessary due to differences in the communication ranges across the platforms that will attempt to adopt the models. For instance, the metric model can be realized with omni-directional antennas, as well as infrared LED sensors. The LED range is considerably smaller (10 cm in Kilobots [28]). Similarly, exploring the effects of different values of n_{top} will be useful. The topological model can be implemented using band-limited communication channels [16], and for infrared-based, band-limited platforms, such as the r-one, n_{top} will be inversely related to the maximum communication range [23].

Acknowledgements This material is based upon research supported by, or in part by, the U.S. Office of Naval Research under award #N000141210987.

References

1. Abaid, N., Porfiri, M.: Fish in a ring: spatio-temporal pattern formation in one-dimensional animal groups. J. R. Soc. Interface **7**(51), 1441–1453 (2010)
2. Aoki, I.: A simulation study on the schooling mechanism in fish. Bull. Japan. Soc. Sci. Fish. **48**(8), 1081–1088 (1982)
3. Ballerini, M., Cabibbo, N., Candelier, R., Cavagna, A., Cisbani, E., Giardina, I., Lecomte, V., Orlandi, A., Parisi, G., Procaccini, A., Viale, M., Zdravkovic, V.: Interaction ruling animal collective behavior depends on topological rather than metric distance: evidence from a field study. PNAS **105**(4), 1232–1237 (2008)

4. Barberis, L., Albano, E.V.: Evidence of a robust universality class in the critical behavior of self-propelled agents: metric versus topological interactions. Phys. Rev. E **89**(1), 9–26 (2014)
5. Bode, N.W.F., Frank, D.W., Wood, A.J.: Limited interactions in flocks: relating model simulations to empirical data. J. R. Soc. Interface **8**, 301–304 (2011)
6. Bonabeau, E., Dorigo, M., Theraulaz, G.: Swarm intelligence: from natural to artificial systems. Oxford University Press, New York, NY (1999)
7. Bonabeau, E., Dorigo, M., Theraulaz, G.: Swarm Intelligence: From Natural to Artificial Systems, No. 1. Oxford University Press (1999)
8. Camazine, S., Deneubourg, J.L., Franks, N.R., Sneyd, J., Theraulaz, G., Bonabeau, E.: Self-Organization in Biological Systems. Princeton University Press, Princeton, NJ (2003)
9. Cianci, C.M., Raemy, X., Pugh, J., Martinoli, A.: Communication in a swarm of miniature robots: the e-puck as an educational tool for swarm robotics. In: Proceedings of the 2nd International Conference on Swarm Robotics, pp. 103–115 (2007)
10. Clough, B.T.: Uav swarming? so what are those swarms, what are the implications, and how do we handle them?. Technical report, DTIC Document (2002)
11. Couzin, I.D., Krause, J., Franks, N.R., Levin, S.A.: Effective leadership and decision-making in animal groups on the move. Nature **433**, 513–516 (2005)
12. Couzin, I.D., Krause, J., James, R., Ruxton, G.D., Franks, N.R.: Collective memory and spatial sorting in animal groups. J. Theor. Biol. **218**(1), 1–11 (2002)
13. Easley, D., Kleinberg, J.: Networks, Crowds, and Markets: Reasoning About a Highly Connected World. Cambridge University Press, New York, NY (2010)
14. Ferrante, E., Turgut, A.E., Mathews, N., Birattari, M., Dorigo, M.: Flocking in stationary and non-stationary environments: a novel communication strategy for heading alignment. In: Parallel Problem Solving from Nature, pp. 331–340 (2010)
15. Goodrich, M., Kerman, S., Pendleton, B., Sujit, P.: What types of interactions do bio-inspired robot swarms and flocks afford a human? In: Proceedings of Robotics: Science and Systems, pp. 105–112 (2012)
16. Goodrich, M.A., Sujit, P., Kerman, S., Pendleton, B., Pinto, J.: Enabling human interaction with bio-inspired robot teams: topologies, leaders, predators, and stakeholders. Technical Report BYU-HCMI 2011-1, Computer Science Department, Brigham Young University, Provo, USA, Aug 2011
17. Huth, A., Wissel, C.: The simulation of the movement of fish schools. J. Theor. Biol. **156**, 365–385 (1992)
18. Kolling, A., Walker, P., Chakraborty, N., Sycara, K., Lewis, M.: Human interaction with robot swarms: a survey. IEEE Trans. Hum.-Mach. Syst. **46**(1), 9–26 (2016)
19. Kolpas, A., Busch, M., Li, H., Couzin, I.D., Petzold, L., Moehlis, J.: How the spatial position of individuals affects their influence on swarms: a numerical comparison of two popular swarm dynamics models. PLoS ONE **8**(3), e58525 (2013)
20. Krause, J., Ruxton, G.D.: Living in Groups. Oxford University Press, New York, NY (2002)
21. Maçãs, C., Cruz, P., Martins, P., Machado, P.: Swarm systems in the visualization of consumption patterns. In: Proceedings of the Twenty-Fourth International Joint Conference on Artificial Intelligence, pp. 2466–2472 (2015)
22. Mannor, S., Shamma, J.: Multi-agent learning for engineers. Artif. Intell. **171**(7), 417–422 (2007)
23. McLurkin, J., McMullen, A., Robbins, N., Habibi, G., Becker, A., Chou, A., Li, H., John, M., Okeke, N., Rykowski, J., Kim, S., Xie, W., Vaughn, T., Zhou, Y., Shen, J., Chen, N., Kaseman, Q., Langford, L., Hunt, J., Boone, A., Koch, K.: A robot system design for low-cost multi-robot manipulation. In: Proceedings of the IEEE/RSJ International Conference on Intelligent Robots and Systems (IROS) (2014)
24. Mondada, F., Gambardella, L.M., Floreano, D., Nolfi, S., Deneubourg, J.L., Dorigo, M.: The cooperation of swarm-bots: physical interactions in collective robotics. Robot. Autom. Mag. **12**(2), 21–28 (2005)
25. Nunamaker, J.F., Applegate, L.M., Konsynski, B.R.: Computer-aided deliberation: model management and group decision support: special focus article. Oper. Res. **36**(6), 826–848 (1988)

26. Parrish, J.K., Viscido, S.V., Grünbaum, D.: Self-organized fish schools: an examination of emergent properties. Biol. Bull. **202**, 296–305 (2002)
27. Reynolds, C.: Flocks, herds and schools: a distributed behavioral model. Comput. Graph. **21**(4), 25–34 (1987)
28. Rubenstein, M., Ahler, C., Nagpal, R.: Kilobot: A low cost scalable robot system for collective behaviors. In: Proceedings of the IEEE International Conference on Robotics and Automation (ICRA) (2012)
29. Shang, Y., Bouffanais, R.: Consensus reaching in swarms ruled by a hybrid metric-topological distance. Eur. Phys. J. B **87**(294) (2014)
30. Shoham, Y., Powers, R., Grenager, T.: If multi-agent learning is the answer, what is the question? Artif. Intell. **171**(7), 365–377 (2007)
31. Strandburg-Peshkin, A., Twomey, C.R., Bode, N.W., Kao, A.B., Katz, Y., Ioannou, C.C., Rosenthal, S.B., Torney, C.J., Wu, H.S., Levin, S.A., Couzin, I.D.: Visual sensory networks and effective information transfer in animal groups. Curr. Biol. **23**(17), R709–R711 (2013)
32. Vicsek, T., Czirók, A., Ben-Jacob, E., Cohen, I., Shochet, O.: Novel type of phase transition in a system of self-driven particles. Phys. Rev. Lett. **75**(6), 1226–1229 (1995)
33. Yanco, H.A., Drury, J.: Classifying human-robot interaction: an updated taxonomy. In: 2004 IEEE International Conference on Systems, Man and Cybernetics, vol. 3, pp. 2841–2846. IEEE (2004)

Target-Dependent Sentiment Analysis of Tweets Using Bidirectional Gated Recurrent Neural Networks

Mohammed Jabreel, Fadi Hassan and Antonio Moreno

Abstract The task of target-dependent sentiment analysis aims to identify the sentiment polarity towards a certain target in a given text. All the existing models of this task assume that the target is known. This fact has motivated us to develop an end-to-end target-dependent sentiment analysis system. To the extent of our knowledge, this is the first system that identifies and extract the target of the tweets. The proposed system is composed of two main steps. First, the targets of the tweet to be analysed are extracted. Afterwards, the system identifies the polarities of the tweet towards each extracted target. We have evaluated the effectiveness of the proposed model on a benchmark dataset from Twitter. The experiments show that our proposed system outperforms the state-of-the-are methods for target-dependent sentiment analysis.

1 Introduction

Sentiment analysis (SA) (also known as opinion mining) is the problem of identifying people's opinions, sentiments or attitudes expressed in text. It normally involves the classification of text into categories such as positive, negative and neutral.

Due to the rapid growth of social networks on the Internet, SA has been applied to analyse opinions on Twitter, Facebook and other digital communities in real time. Sentiment analysis has now a wide range of applications in fields like marketing, management, e-health, politics and tourism [10, 11, 19]. For instance, it can enhance the capabilities of customer relationship management systems and recommenders

M. Jabreel (✉) · F. Hassan · A. Moreno
Intelligent Technologies for Advanced Knowledge Acquisition (ITAKA),
Departament d'Enginyeria Informàtica i Matemàtiques, Universitat Rovira i Virgili,
Av. Països Catalans, 26, 43007 Tarragona, Spain
e-mail: mohammed.jabreel@urv.cat

F. Hassan
e-mail: fadi.abdulfattahmohammed@estudiants.urv.cat

A. Moreno
e-mail: antonio.moreno@urv.cat

I. Hatzilygeroudis and V. Palade (eds.), *Advances in Hybridization of Intelligent Methods*, Smart Innovation, Systems and Technologies 85,
https://doi.org/10.1007/978-3-319-66790-4_3

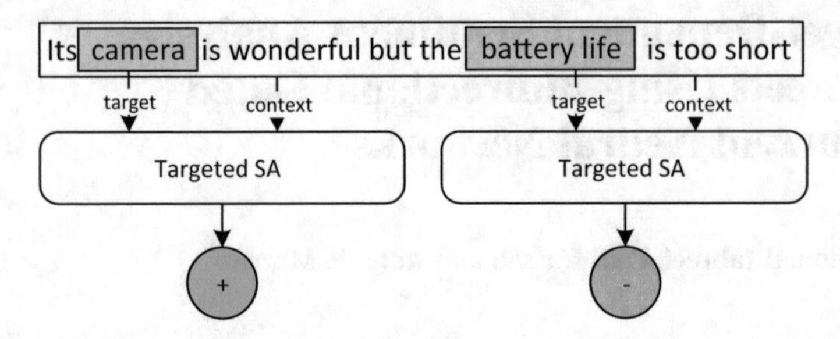

Fig. 1 Example of the two steps. The filled rectangle represents the target extraction step, where the rounded rectangle represents the sentiment analysis step and it receives two inputs the extracted target and its context

by finding out which features customers are particularly interested in or avoiding the recommendation of items that have received unfavourable feedbacks.

SA can be done at different levels. Coarse-grained analysis attempt to extract the overall polarity on a document or sentence level, whereas, in a fine-grained level of analysis, the problem is to identify the sentiment polarity towards a certain target in a given text (*Target-dependent sentiment analysis*) [3, 13, 35]. In this problem it is necessary to determine the target and its context, which can be defined as follows:

Target A *target* is an entity (person, organisation, product, object, etc.) referred to in a text, about which an opinion is expressed.

Context The *context* of the target is the text surrounding it, that provides information about the polarity of the sentiment towards it.

It is quite usual to give several opinions on different aspects of an object in a single sentence. For example, the text "*I have got a new mobile. Its camera is wonderful but the battery life is too short.*", gives both positive and negative remarks about a mobile phone. It may be seen that the example contains three targets ("*mobile*", "*camera*" and "*battery life*") and the sentiment polarities towards them can be seen as "neutral", "positive" and "negative", respectively. Such fine-grained opinions are important for both producers and customers [22]. The importance of target information has been proven by previous studies. It has been shown [13] that about 40% of the errors of sentiment analysis systems are caused by the lack of information about the target. Thus, the target-dependent SA problem can be addressed by designing a system with two steps, shown in Fig. 1. The first step aims to extract or identify the target in a given text, while the objective of the second step is to identify the opinion expressed in the text towards the extracted target. Those steps are commented in the following subsections.

1.1 Target Identification

Target-dependent sentiment analysis on Twitter is the problem of identifying the sentiment polarity towards a certain target in a given tweet. Extracting the targets from the tweets is the key task in this problem. However, all the existing studies of this task assume that the target is known. Thus, we have developed a system to identify automatically the explicit targets of the tweets.

Recently, a similar problem to target identification, known as *aspect term extraction*, has been studied extensively. There are two main kinds of approaches: *supervised* and *unsupervised*. In the supervised approaches machine-learning systems are trained on manually annotated data to extract targets in the reviews. The most common techniques employed in supervised approaches are decision trees, support vector machines, K-nearest neighbour, Naive Bayesian classifiers and neural networks [14, 34]. On the other hand, unsupervised approaches aim to automatically extract product features using syntactic and contextual patterns without the need of annotated data [21, 22].

There is one particularly interesting supervised approach, which conceptualizes the aspect extraction problem as a sequence labeling problem [12]. The most successful sequence labeling systems are probabilistic graphical models such as Hidden Markov Models and Conditional Random Fields [16, 29]. However, their main drawback is that they rely heavily on a set of hand-crafted features, whose definition is very time consuming task. Recently, deep neural networks have been utilized to extract automatically high-level features in many tasks such as speech recognition [5], text classification [15], image classification [7], etc. Recurrent Neural Networks (RNNs) have been proved to be a very useful technique to represent sequential data such as text. These models have also shown great success in solving sequence labeling tasks, e.g. Named Entity Recognition and Part-Of-Speech tagging [17, 18]. Following these approaches we propose to use a bidirectional gated recurrent neural network to solve the problem of target extraction, as described in Sect. 3.1.

1.2 Target-Dependent Sentiment Analysis

Most of the current studies on sentiment analysis are inspired by the work presented in [26]. Machine learning techniques have been used to build a classifier from a set of sentences with a manually annotated sentiment polarity. The success of these models is based on two main facts: the availability of a large amount of labeled data and the intelligent manual design of a set of features that can be used to differentiate the samples.

Their performance basically depends on defining an appropriate set of efficient classifying features [4, 20, 25, 28]. For instance, the authors in [24] and [10] used diverse sentiment lexicons and a variety of hand-crafted features in their sentiment analysis systems.

Target-dependent sentiment analysis is also regarded as a text classification problem in the literature. Standard text classification approaches such as feature-based Support Vector Machines [13, 26] can be used to build a sentiment classifier. For instance, the work presented in [13] combined manually designed target-independent features and target-dependent features with expert knowledge, a syntactic parser and external resources.

Recent studies, such as the works proposed by [3, 33, 35, 36], use neural network methods and encode each sentence in a continuous and low-dimensional vector space without feature engineering. Dong et al. [3] transformed a sentence dependency tree into a target-specific recursive structure, and used an *Adaptive Recursive Neural Network* to learn a higher level representation. Vo and Zhang [35] used rich features including sentiment-specific word embedding and sentiment lexicons. The work presented in [36] modeled the interaction between the target and the surrounding context using a gated neural network. Tang et al. [33] developed long short-term memory models to capture the relatedness of a target word with its context words when composing the continuous representation of a sentence. Most of these studies rely on the idea of splitting the sentence/text into target, left context and right context.

Unlike previous studies, we propose a *target-dependent bidirectional gated recurrent unit* (TD-biGRU), which is capable of modeling the relatedness between target words and their contexts by concatenating an embedded vector that represents the target word(s) with two vectors that capture both the preceding and following contextual information. Section 3.2 describes the proposed model in detail.

The rest of this chapter is structured as follows. Section 2 presents the basic concepts used in this work. In Sect. 3 the proposed models are described. The experiments and results are presented and discussed in Sect. 4. Finally, in the last section the conclusions and lines of future work are outlined.

2 Background

This section explains briefly the basic concepts used in this work. We start by explaining the vector representations of words, and then we describe recurrent neural networks, gated recurrent units, bidirectional recurrent neural networks and the softmax classifier.

2.1 Vector Representations of Words

Word embeddings are an approach for distributional semantics which represents words as vectors of real numbers. Such representation has useful clustering properties, since the words that are semantically and syntactically related are represented by similar vectors [23]. For example, the words "coffee" and "tea" will be very close in the created space.

When a text has to be analysed, the first step is to map each word into a continuous, low dimensional and real-valued vector, which can later be processed by a neural network model. All the word vectors are stacked into a matrix $E \in \mathbb{R}^{d \times N}$, where N is the vocabulary size and d is the vector dimension. This matrix is called the *embedding layer* or the *lookup table layer*. The embedding matrix can be initialized using a pre-trained model like *word2vec* or *Glove* [23, 27]. In this work, the embedding layer contains a pre-trained model which was learned using the *Glove* algorithm [27] on a large corpus of two billions of tweets (short textual messages sent through the Twitter social network).

2.2 Recurrent Neural Networks (RNNs)

A Recurrent Neural Network (RNN) is a type of neural network architecture specifically designed for modeling sequential inputs of varying lengths such as text.

As shown in Fig. 2, at each time step t, it takes the input vector $x \in \mathbb{R}^d$ and the hidden state vector $h_{t-1} \in \mathbb{R}^{d_h}$ and outputs the next hidden state h_t by applying the following equation:

$$h_t = \phi\left(x_t, h_{t-1}\right) \tag{1}$$

Usually, h_0 is initialized to a zero vector in order to calculate the first hidden state.

The most common approach is to use the affine transformation operation followed by an element-wise non-linearity, e.g. Rectified Linear Unit (ReLU), as the function ϕ that produces the next hidden state vector h_t.

$$\phi(x_t, h_{t-1}) = f(Wx_t + Vh_{t-1} + b) \tag{2}$$

In this formula, $W \in \mathbb{R}^{d \times d_h}$, $V \in \mathbb{R}^{d_h \times d_h}$ and $b \in \mathbb{R}^{d_h}$ are the parameters of the model, and f is an element-wise non-linearity.

In practice, the major issue of RNNs using these transition functions is the difficulty of learning long-term dependencies due to vanishing/exploding gradients [1].

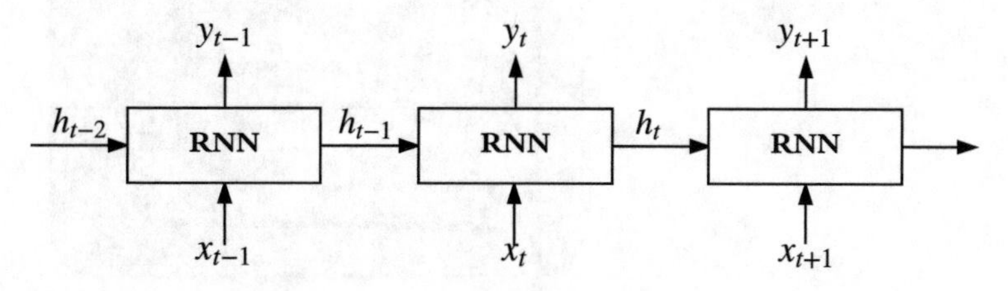

Fig. 2 Recurrent Neural Network

Long short-term memory (LSTM) units [9] and Gated Recurrent Unit (GRU) [2] have been specifically designed to address this problem. In this work we use a GRU as ϕ, and we explain how it is used to produce the hidden state vector h_t in the next subsection.

2.3 Gated Recurrent Unit (GRU)

Gated recurrent units (GRUs) were designed to have more persistent memory, making them very useful to capture long-term dependencies between the elements of a sequence. GRUs are the basic components of the model proposed in this chapter. Figure 3 shows a graphical depiction of a gated recurrent unit.

This kind of units have *reset* (r_t) and *update* (z_t) gates. The former has the ability to completely reduce the past hidden state h_{t-1} if it considers that it is irrelevant to the computation of the new state, whereas the later is responsible for determining how much of h_{t-1} should be carried forward to the next state h_t.

The output h_t of a GRU depends on the input x_t and the previous state h_{t-1}, and it is computed as follows:

$$r_t = \sigma\left(W_r \cdot [h_{t-1}; x_t] + b_r\right) \tag{3}$$

$$z_t = \sigma\left(W_z \cdot [h_{t-1}; x_t] + b_z\right) \tag{4}$$

$$\widetilde{h}_t = tanh\left(W_h \cdot [(r_t \odot h_{t-1}); x_t] + b_h\right) \tag{5}$$

$$h_t = (1 - z_t) \odot h_{t-1} + z_t \odot \widetilde{h}_t \tag{6}$$

In these expressions r_t and z_t denote the *reset* and *update* gates, $\widetilde{h}_t$ is the candidate output state and h_t is the actual output state at time t. The symbol $\odot$ stands for element-wise multiplication, σ is a sigmoid function and ; stands for the vector-

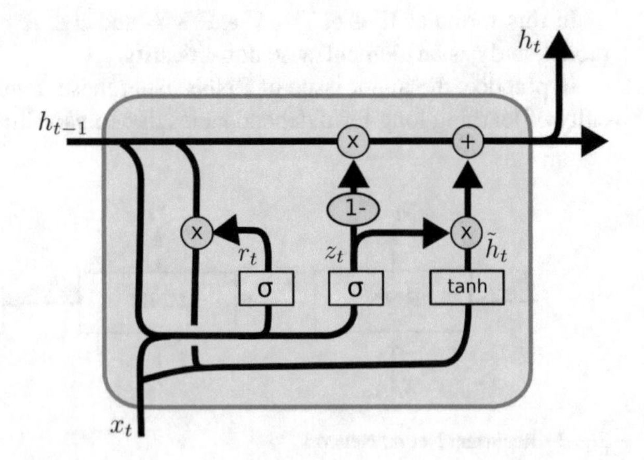

Fig. 3 Gated Recurrent Unit (GRU). *Figure source* http://www.colah.github.io/posts/2015-08-Understanding-LSTMs/

concatenation operation. $W_r, W_z, W_h \in \mathbb{R}^{d_h \times (d+d_h)}$ and $b_r, b_z, b_h \in \mathbb{R}^{d_h}$ are the parameters of the *reset* and *update* gates, where d_h is the dimension of the hidden state.

2.4 Bidirectional RNNs

The standard RNN, described in Sect. 2.2, reads an input sequence $X = (x_1, \dots, x_n)$ in a forward direction (left-to-right) starting from the first symbol x_1 and ending in the last one x_n. Thus, it processes sequences in temporal order, ignoring the future context. For many tasks on sequences it is beneficial to have access to future as well as to past information. For example, in text processing, decisions are usually made after the whole sentence is known. The Bidirectional BiRNN architecture [6] proposed a solution for making predictions based on both past and future information.

Figure 4 illustrates the architecture of a BiRNN, it consists of forward $\vec{\phi}$ and backward $\overleftarrow{\phi}$ RNNs. The first one reads the input sequence in a forward direction $(x_1, \dots, x_n)$ and produces a sequence of forward hidden states $(\overrightarrow{h_1}, \dots, \overrightarrow{h_n})$, whereas the former reads the sequence in the reverse order $(x_n, \dots, x_1)$ resulting in a sequence of backward hidden states $(\overleftarrow{h_n}, \dots, \overleftarrow{h_1})$.

We obtain a representation for each word x_t by concatenating the corresponding forward hidden state $\overrightarrow{h_t}$ and the backward one $\overleftarrow{h_t}$. The following equations illustrate the main ideas:

$$\overrightarrow{h_t} = \vec{\phi}(x_t, \overrightarrow{h_{t-1}}) \tag{7}$$

$$\overleftarrow{h_t} = \overleftarrow{\phi}(x_t, \overleftarrow{h_{t-1}}) \tag{8}$$

$$h_t = [\overrightarrow{h_t}; \overleftarrow{h_t}] \tag{9}$$

In this work we use two GRUs, one as $\vec{\phi}$ and the other as $\overleftarrow{\phi}$. We call this model biGRU.

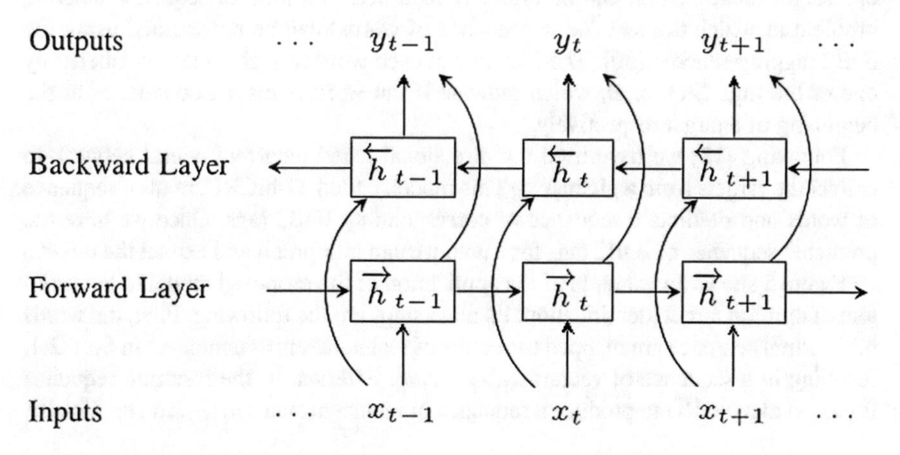

Fig. 4 Bidirectional Recurrent Neural Network

2.5 Softmax Classifier

The softmax classifier is a feed-forward neural network followed by the softmax function, which is used for multi-class classification (under the assumption that the classes are mutually exclusive). It takes as input a vector $v \in \mathbb{R}^m$ and produces the probabilities for each class as follows:

$$p(y = i|v; W, b) = \frac{exp(w_i^T v + b_i)}{\sum_{j=1}^{C} exp(w_j^T v + b_j)}, i = 1, 2, \ldots, C \quad (10)$$

This can be interpreted as the (normalized) probability assigned to each class i given the input vector v, and parameterized by $W \in \mathbb{R}^{m \times C}$ and $b \in \mathbb{R}^C$, where C is the number of classes, w_i is the i-th column of W and b_i is a bias term.

3 Model Description

We describe in this section the proposed model to tackle the problem of target-dependent SA. It is composed of two main steps. First, the target of the tweet to be analysed is identified as described in next subsection. Once the target has been obtained, it is passed together with the tweet as input to the model described in Sect. 3.2 to determine the sentiment polarity.

3.1 Target Identification

In this step we aim to extract the targets which customers expressed their opinions on. Target identification can be typically regarded as a kind of sequence labeling problem in which the text (i.e. a sequence of words) can be represented using the IOB2 tagging scheme [30]. The idea is that each word in a given text is labeled by one of the tags *I*, *O*, or *B*, which indicate if the word is inside, outside, or at the beginning of a target respectively.

Following [12] we have used a bidirectional gated recurrent neural network to extract the targets from a given text. This model, called TI-biGRU, reads a sequence of words and predicts a sequence of corresponding IOB2 tags. Once we have the predicted sequence of IOB2 tags for a text, we can interpret it and extract the targets.

Figure 5 shows an example of the application of the proposed model to the problem of opinion target identification. Its main steps are the following. First, the words of the input sentence are mapped to vectors of real numbers as explained in Sect. 2.1, resulting in a sequence of vectors $x_1, x_2, \ldots, x_n$. Afterwards, the resulting sequence is passed to a biGRU to produce a sequence of recurrent states $h_1, h_2, \ldots, h_n$. Finally,

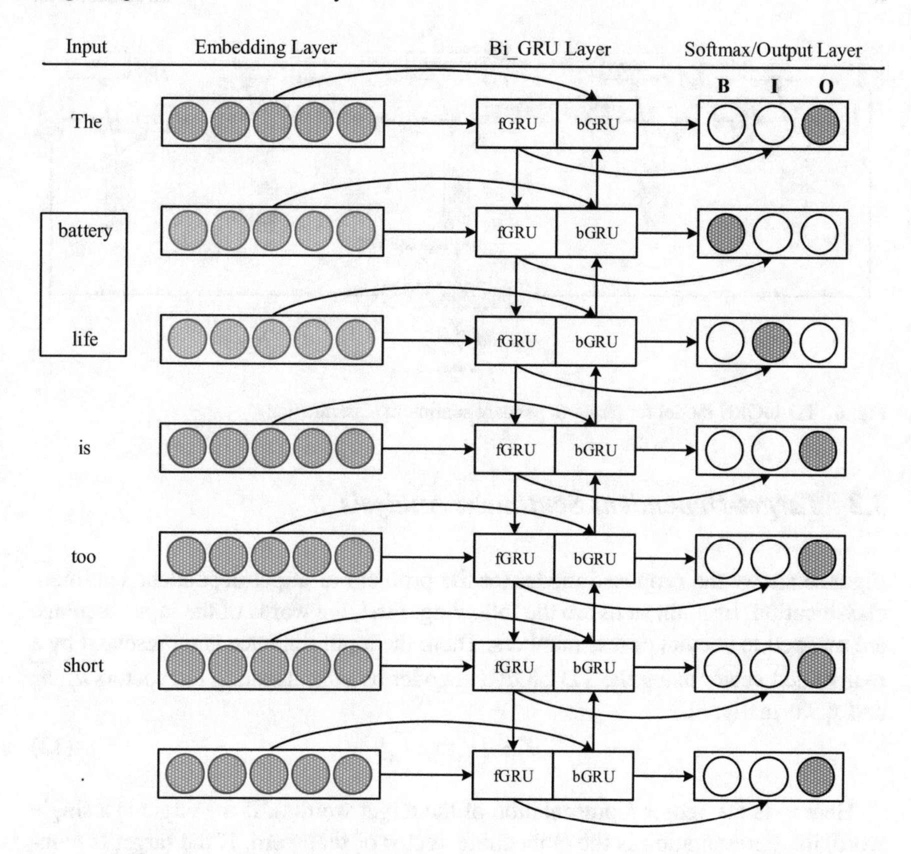

Fig. 5 TI-biGRU model for target identification

each produced sequence element h_i is passed through a softmax layer to predict the probability distribution over the three possible output tags (I, O or B).

The model is trained to minimize the following objective function, which is the cross-entropy between the expected tag and the predicted tag distribution of each word i:

$$J = -\sum_{s\in S}\sum_{i=1}^{n}\sum_{t=1}^{3} p_i^s(t) log(P(y = t|h_i^s)) \quad (11)$$

In this expression $p_i^s(t) \in \{0, 1\}$ is the ground-truth function which indicates whether tag t is the correct tag for the word i in the sentence s and S is the set of the sentences in the training set. The derivative of the objective function J is taken through back-propagation with respect to the whole set of parameters of the model. These parameters are optimized using the stochastic optimization method *RMSProp*.

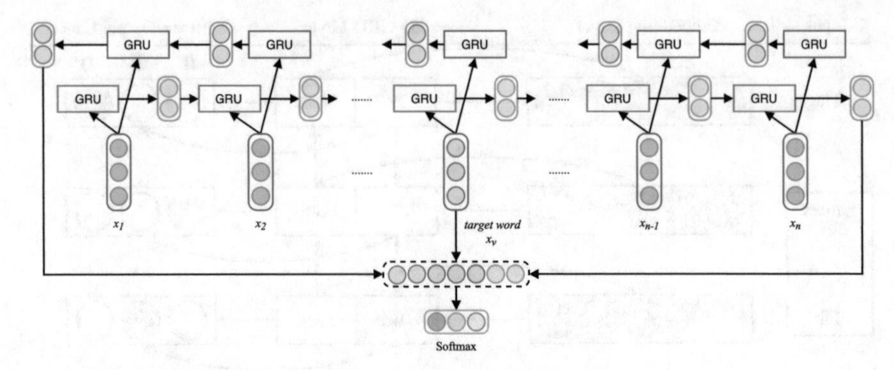

Fig. 6 TD-biGRU model for target-dependent sentiment classification

3.2 Target-Dependent Sentiment Analysis

Figure 6 shows the proposed model for the problem of target-dependent sentiment classification. Its main steps are the following. First, the words of the input sentence are mapped to vectors of real numbers. Then, the input sentence is represented by a real-valued vector using the TD-biGRU encoder by concatenating the vectors $\overrightarrow{h_n}, \overleftarrow{h_n}$ and x_v, formally:

$$X = [\overrightarrow{h_n}; x_v; \overleftarrow{h_n}] \tag{12}$$

Here x_v is the vector representation of the target word(s). If the target is a single word, its representation is the embedding vector of that word. If the target is composed of multiple words, such as "battery life", its representation is the average of the embedding vectors of the words [32].

In this way, the obtained vector summarizes the input sentence and contains semantic, syntactic and/or sentimental information based on the word vectors. Finally, this vector is passed through a softmax classifier to classify the sentence into positive, negative or neutral.

We trained the system to minimize the following categorical cross-entropy:

$$J = -\sum_{s \in S} \sum_{c=1}^{3} G_c(s) log(P(y = c|s)) \tag{13}$$

In this expression S is the training set and $G_c(s) \in \{0, 1\}$ is the ground-truth function which indicates whether class c is the correct sentiment category for sentence s.

The derivative of the objective function is taken through back-propagation with respect to the whole set of parameters of the model, and these parameters are updated with the stochastic gradient descent. The learning rate is initially set to 0.1 and the parameters are initialized randomly over a uniform distribution in $[-0.03, 0.03]$. For the regularization, dropout layers [8, 31] are used with probability 0.5 on the lookup-table output to the GRU input and on the concatenation output to the softmax input.

4 Experiments and Results

4.1 Datasets

We evaluated the effectiveness of the proposed models by using them in the supervised tasks of target identification and target-dependent sentiment classification on the benchmark dataset provided in [3]. The dataset contains 6248 training examples and 692 examples in the testing set. Each example in the dataset contains the sentence, the target and the label of sentiment polarity. In case the sentence contains more than one target with different polarities it is repeated with each one. The numerical description of the positive, negative and neutral examples is shown in Table 1.

4.2 Evaluation Metrics

The evaluation metrics of the target identification problem are the precision (the number of correct targets divided by the number of all returned targets), recall (the number of correct targets divided by the number of targets that should have been returned) and F_1 (the harmonic mean of precision and recall), which can be defined as follows:

$$Precision = \frac{|S \cap G|}{|S|} \tag{14}$$

$$Recall = \frac{|S \cap G|}{|G|} \tag{15}$$

$$F_1 = \frac{2 \cdot Precision \cdot Recall}{Precision + Recall} \tag{16}$$

Here S is the set of the predicted targets that the system returned for all the test examples, and G is the set of the gold (correct) targets.

Table 1 Numerical description of the dataset

	Training	Testing	Percentage (%)
#Positives	1562	173	25
#Neutrals	3124	346	50
#Negatives	1562	173	25
Total	6248	692	

The evaluation metrics of the target-dependent sentiment analysis system are the classification accuracy (the percentage of examples that are correctly classified) and the Macro-F1 measure (the averaged F1 measure over the three sentiment classes).

4.3 Results and Discussions

4.3.1 Target Identification

As stated before, all the existing models of target-dependent SA assume that the target is known. Thus, to the extent of our knowledge, this is the first target-dependent SA system that identifies and extracts the target of the tweets. The typical RNN model defined in Eq. (2) is used as our baseline. We investigated the effectiveness of TI-biGRU, which is used to automatically identify the target from a tweet, by comparing it with the baseline model.

In Table 2 baseline-I and baseline-II denote the typical RNN and biRNN respectively, while TI-GRU is the simplified version of TI-biGRU in which only the past information is considered, ignoring the bidirectionality. It is clearly shown that TI-biGRU outperforms the other models. Another interesting observation from the reported result is that both baseline-II and TI-biGRU perform better than their relaxed versions (i.e. baseline-I and TI-GRU). Such conclusion confirms the effectiveness of BiRNNs in this kind of tasks.

From the table, we observe that both TI-GRU and TI-iGRU perform better than the baselines (i.e. the standard RNN based models). In term of recall all the models give interesting results. This can be attributed to the fact that achieving a recall of 100% is trivial by assuming that all words in the sentence/tweet are targets. Therefore, recall alone is not enough and it is also necessary to measure the number of incorrect returned targets by computing the precision.

In addition, we can see that there are remarkable improvements in term of precision. For example, TI-biGRU's precision score is the best and it has 12.8%, 9.57% and 7.65% precision improvements compared with those of baseline-I, baseline-II and TI-GRU, respectively. On the other hand, there are smaller improvements in terms of recall.

Table 2 Comparison of our model to the baselines on target identification. Best scores are shown in bold

Model	Precision	Recall	F_1
baseline-I	77.90	87.57	82.44
baseline-II	79.76	90.17	84.67
TI-GRU	81.18	90.89	86.10
TI-biGRU	**87.39**	**91.18**	**89.25**

4.3.2 Target-Dependent Sentiment Analysis

We compared the proposed model with the state-of-the-art methods used in the task of target-dependent sentiment classification, including:

- **SVM-indep**: Support Vector Machine classifier built with target-independent features, such as unigram, bigram, punctuations, emoticons, hashtags and the numbers of positive or negative words in the General Inquirer sentiment lexicon [13].
- **SVM-dep**: SVM-indep model extended by adding a set of features that represent the target [13].
- **Recursive RNN**: a recursive neural network is employed to learn the feature representation of the examples over a transferred target-dependent dependency tree [3].
- **AdaRNN**: extension of the recursive RNN which uses more than one composition function and adaptively selects them according to the input [3]. AdaRNN has three variations: AdaRNN-w/oE, AdaRNN-w/E and AdaRNN-comb. Unlike AdaRNN-w/oE, AddRNN-w/E model uses the dependency type in the process of composition function selection. AddaRNN-comb combines the root vectors obtained by AdaRNN-w/E with the unigram and bigram features, and then they are fed into a SVM classifier.
- **Target-ind/Target-dep**: SVM classifiers based on a rich set of target-independent and target-dependent features [35]. This model has an extension, called **Target-dep+**, in which sentiment lexicon features have been incorporated.
- **LSTM**, **TD-LSTM**, **TC-LSTM**: these methods are based on the *long short-term memory* model proposed by [33]. In the LSTM model the target is ignored. The idea behind TD-LSTM is to use two LSTM neural networks, so that the left one represents the preceding context plus the target and the right one represents the target plus the following context. TC-LSTM is an extension of TD-LSTM in which a vector that represents the target is concatenated to each context word.

The values under the section “A” in Table 3 represent the results of the baseline model (basic bidirectional gated recurrent units—biGRU—without incorporating target information), the new TD-biGRU model in case the targets are manually given and the results when we apply the two steps of our system to analyse the tweets. Each tweet is passed to the system to first extract the targets and then identify the sentiment polarities towards these targets. Section “B” contains the results of the compared models (obtained from their associated papers). With the exception of AdaRNN, each approach presented in Table 3 has a target-independent version (which does not incorporate any information about targets) and two or three target-dependent versions. For instance, in our case biGRU is the target-independent version.

As it can be observed from the reported results, the target-independent models (SVM-indep, Target-indep, LSTM and biGRU) have a worst performance than the corresponding models that consider the target information (SVM-dep, Target-dep*, TD-LSTM, TC-LSTM and TD-biGRU). This conclusion confirms the fact that ignoring the target information causes about 40% of sentiment analysis errors [13]. It

Table 3 Comparison of different methods on target-dependent sentiment classification. Evaluation metrics are accuracy and macro-F1. Best scores are shown in bold

Model	Accuracy	Macro-F1
A. Our model		
biGRU	69.94	68.40
TD-biGRU	**72.25**	**70.47**
End-To-End-TD	70.08	68.22
B. State-of-the-art systems		
SVM-indep	62.70	60.20
SVM-dep	63.40	63.30
Recursive NN	63.00	62.80
AdaRNN-w/oE	64.90	64.44
AdaRNN-w/E	65.80	65.50
AdaRNN-comb	66.30	65.90
Target-ind	67.30	66.40
Target-dep	69.70	68.00
Target-dep$^+$	71.10	69.90
LSTM	66.50	64.70
TD-LSTM	70.80	69.00
TC-LSTM	71.50	69.50

may also be notived that neural-based models perform better than the feature-based SVM classifiers.

The novel TD-biGRU model outperforms the state-of-the-art models both in terms of accuracy and Macro-F1. Our end-to-end approach gives a comparable results to those models, including our TD-biGRU model, that assume the target is known.

To get more insight on this result, we analyzed the confusion matrix given by the TD-biGRU model to figure out which are the most common incorrect cases. Figure 7 shows the confusion matrix obtained by applying TD-biGRU. As observed, the matching between the true and the predicted labels is quite high (matrix diagonal). Out of the 192 misclassified samples, 76 (39.6%) of them were misclassified between negative and neutral (i.e., either negative samples were misclassified as neutral or viceversa) and 31 (16.1%) samples were misclassified between negative and positive. The number of samples misclassified between positive and neutral is 85 (44.3%).

This analysis shows that most of the misclassified examples are related to the neutral category. We believe that this problem can be handled by adding more information (e.g. lexicon information). We leave the study of this hypothesis for the future work.

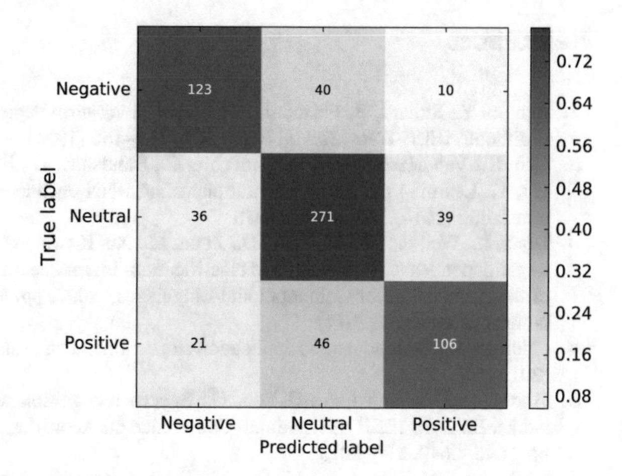

Fig. 7 Confusion matrix

5 Conclusion

We have developed an end-to-end target-dependent Twitter sentiment analysis system. The proposed model has the ability of identifying and extracting the target of the tweets, representing the relatedness between the targets and its contexts and identifying the polarities of the tweets towards the targets. The effectiveness of the proposed system has been evaluated on a benchmark of tweets, obtaining results that outperform the state-of-the-art models. The confusion matrix of the results obtained by TD-biGRU shows that most of the misclassified examples are related to the neutral category.

In the future work we plan to extend our system to handle this weakness by integrating more information such as lexicon information and/or the dependency tree. Our system extracts only the targets that are mentioned explicitly in the tweets. However, it is sometimes recognized that targets are mentioned implicitly in tweets and they are detected from the context. Thus, we will consider this point in our future work, by designing a system that can detect both the explicit targets and the implicit targets that are not mentioned in the tweets. Although joint learning of all subsystems has been proved to be useful in natural language processing and text analysis tasks, in this work we have trained each subsystem (i.e. the target identification and the targeted SA) independently and we have combined them in the inference step. Thus, we plan to extend our system and apply this learning technique.

Acknowledgements The authors acknowledge the support of Univ. Rovira i Virgili through a Martí i Franqués Ph.D. grant, the assistant/teaching grant for the Department of Computer Engineering and Mathematics and the Research Support Funds 2016PFR-URV-B2-60.

References

1. Bengio, Y., Simard, P., Frasconi, P.: Learning long-term dependencies with gradient descent is difficult. IEEE Trans. Neural Netw. **5**(2), 157–166 (1994)
2. Cho, K., Van Merriënboer, B., Gulcehre, C., Bahdanau, D., Bougares, F., Schwenk H., Bengio, Y.: Learning phrase representations using RNN encoder-decoder for statistical machine translation (2014). arXiv:1406.1078
3. Dong, L., Wei, F., Tan, C., Tang, D., Zhou, M., Xu, K.: Adaptive recursive neural network for target-dependent twitter sentiment classification. In: Proceedings of the 52nd Annual Meeting of the Association for Computational Linguistics, vol. 2, pp. 49–54. Association for Computational Linguistics (2014)
4. Feldman, R.: Techniques and applications for sentiment analysis. Commun. ACM **56**(4), 82–89 (2013)
5. Graves, A., Mohamed, A., Hinton, G.: Speech recognition with deep recurrent neural networks. In: 2013 IEEE International Conference on Acoustics, Speech and Signal Processing, pp. 6645–6649, May 2013
6. Graves, A., Mohamed, A., Hinton, G.: Speech recognition with deep recurrent neural networks. In: 2013 IEEE International Conference on Acoustics, Speech and Signal Processing (ICASSP), pp. 6645–6649. IEEE (2013)
7. He, K., Zhang, X., Ren, S., Sun, J.: Deep residual learning for image recognition. In: 2016 IEEE Conference on Computer Vision and Pattern Recognition (CVPR), pp. 770–778, June 2016
8. Hinton, G.E., Srivastava, N., Krizhevsky, A., Sutskever, I., Salakhutdinov, R.R.: Improving neural networks by preventing co-adaptation of feature detectors (2012). arXiv:1207.0580
9. Hochreiter, S., Schmidhuber, J.: Long short-term memory. Neural Comput. **9**(8), 1735–1780 (1997)
10. Jabreel, M., Moreno, A.: Sentirich: sentiment analysis of tweets based on a rich set of features. In: Artificial Intelligence Research and Development—Proceedings of the 19th International Conference of the Catalan Association for Artificial Intelligence, Barcelona, Catalonia, Spain, 19–21 Oct 2016, pp. 137–146 (2016)
11. Jabreel, M., Moreno, A., Huertas, A.: Do local residents and visitors express the same sentiments on destinations through social media? In: Information and Communication Technologies in Tourism 2017, pp. 655–668. Springer (2017)
12. Jebbara, S., Cimiano, P.: Aspect-based sentiment analysis using a two-step neural network architecture. In: Semantic Web Evaluation Challenge, pp. 153–167. Springer (2016)
13. Jiang, L., Yu, M., Zhou, M., Liu, X., Zhao, T.: Target-dependent twitter sentiment classification. In: Proceedings of the 49th Annual Meeting of the Association for Computational Linguistics: Human Language Technologies, vol. 1, pp. 151–160. Association for Computational Linguistics (2011)
14. Kessler, J.S., Nicolov, N.: Targeting sentiment expressions through supervised ranking of linguistic configurations. In: ICWSM (2009)
15. Kim, Y.: Convolutional neural networks for sentence classification (2014). arXiv:1408.5882
16. Lafferty, J., McCallum, A., Pereira, F., et al.: Conditional random fields: probabilistic models for segmenting and labeling sequence data. In: Proceedings of the Eighteenth International Conference on Machine Learning, ICML, vol. 1, pp. 282–289 (2001)
17. Lample, G., Ballesteros, M., Subramanian, S., Kawakami, K., Dyer, C.: Neural architectures for named entity recognition (2016). arXiv:1603.01360
18. Ling, W., Dyer, C., Black, A.W., Trancoso, I., Fernandez, R., Amir, S., Marujo, L., Luis, T.: Finding function in form: compositional character models for open vocabulary word representation. In: Proceedings of the 2015 Conference on Empirical Methods in Natural Language Processing, pp. 1520–1530, Lisbon, Portugal, Sept 2015. Association for Computational Linguistics
19. Liu, B.: Opinion mining and sentiment analysis. In: Web Data Mining, pp. 459–526. Springer (2011)

20. Liu, B.: Sentiment analysis and opinion mining. In: Synthesis lectures on Human Language Technologies, vol. 5, no. 1, pp. 1–167 (2012)
21. Liu, Q., Gao, Z., Liu, B., Zhang, Y.: Automated rule selection for aspect extraction in opinion mining. In: IJCAI, pp. 1291–1297 (2015)
22. Liu, Q., Gao, Z., Liu, B., Zhang, Y.: Automated rule selection for opinion target extraction. Knowl.-Based Syst. **104**, 74–88 (2016)
23. Mikolov, T., Chen, K., Corrado, G., Dean, J.: Efficient estimation of word representations in vector space (2013). arXiv:1301.3781
24. Mohammad, S., Kiritchenko, S., Zhu, X.: NRC-Canada: building the state-of-the-art in sentiment analysis of tweets. In: Proceedings of the Seventh International Workshop on Semantic Evaluation Exercises (SemEval-2013), Atlanta, Georgia, USA, June 2013
25. Pang, B., Lee, L.: Opinion mining and sentiment analysis. Found. Trends Inf. Retr. **2**(1–2), 1–135 (2008)
26. Pang, B., Lee, L., Vaithyanathan, S.: Thumbs up?: sentiment classification using machine learning techniques. In: Proceedings of the ACL-02 Conference on Empirical Methods in Natural Language Processing, EMNLP'02, vol. 10, pp. 79–86, Stroudsburg, PA, USA. Association for Computational Linguistics (2002)
27. Pennington, J., Socher, R., Manning, C.D.: Glove: global vectors for word representation. In: Empirical Methods in Natural Language Processing (EMNLP), pp. 1532–1543 (2014)
28. Perikos, I., Hatzilygeroudis, I.: Recognizing emotions in text using ensemble of classifiers. Eng. Appl. Artif. Intell. **51**, 191–201 (2016)
29. Ratinov, L., Roth, D.: Design challenges and misconceptions in named entity recognition. In: Proceedings of the Thirteenth Conference on Computational Natural Language Learning, pp. 147–155. Association for Computational Linguistics (2009)
30. Sang, E.F., Veenstra, J.: Representing text chunks. In: Proceedings of the Ninth Conference on European Chapter of the Association for Computational Linguistics, pp. 173–179. Association for Computational Linguistics (1999)
31. Srivastava, N., Hinton, G., Krizhevsky, A., Sutskever, I., Salakhutdinov, R.: Dropout: a simple way to prevent neural networks from overfitting. J. Mach. Learn. Res. **15**, 1929–1958 (2014)
32. Sun, Y., Lin, L., Tang, D., Yang, N., Ji, Z., Wang, X.: Modeling mention, context and entity with neural networks for entity disambiguation. In: Proceedings of the 24th International Conference on Artificial Intelligence, IJCAI'15, pp. 1333–1339. AAAI Press (2015)
33. Tang, D., Qin, B., Feng, X., Liu, T., Target-dependent sentiment classification with long short term memory (2015). arXiv:1512.01100
34. Toh, Z., Su, J.: Nlangp at semeval-2016 task 5: improving aspect based sentiment analysis using neural network features. In: Proceedings of SemEval, pp. 282–288 (2016)
35. Vo, D.-T., Zhang, Y.: Target-dependent twitter sentiment classification with rich automatic features. In: Proceedings of the Twenty-Fourth International Joint Conference on Artificial Intelligence (IJCAI 2015), pp. 1347–1353 (2015)
36. Zhang, M., Zhang, Y., Vo, D.-T.: Gated neural networks for targeted sentiment analysis. In: Proceedings of the Thirtieth AAAI Conference on Artificial Intelligence, Phoenix, Arizona, USA, pp. 3087–3093. Association for the Advancement of Artificial Intelligence (2016)

Traffic Modelling, Visualisation and Prediction for Urban Mobility Management

Tomasz Maniak, Rahat Iqbal and Faiyaz Doctor

Abstract Smart city combines connected services from different disciplines offering a promise of increased efficiency in transport and mobility in urban environment. This has been enabled through many important advancements in fields like machine learning, big data analytics, hardware manufacturing and communication technology. Especially important in this context is big data which is fueling the digital revolution in an increasingly knowledge driven society by offering intelligence solutions for the smart city. In this paper, we discuss the importance of big data analytics and computational intelligence techniques for the problem of taxi traffic modelling, visualisation and prediction. This work provides a comprehensive survey of computational intelligence techniques appropriate for the effective processing and analysis of big data. A brief description of many smart city projects, initiatives and challenges in the UK is also presented. We present a hybrid data modelling approach used for the modelling and prediction of taxi usage. The approach introduces a novel biologically inspired universal generative modelling technique called Hierarchical Spatial-Temporal State Machine (HSTSM). The HSTSM modelling approach incorporates many soft computing techniques including: deep belief networks, auto-encoders, agglomerative hierarchical clustering and temporal sequence processing. A case study for the modelling and prediction of traffic based on taxi movements is described, where HSTSM is used to address the computational challenges arising from analysing and processing large volumes of varied data.

T. Maniak
Interactive Coventry Ltd, Coventry University Technology Park, Coventry, UK
e-mail: tomasz.maniok@interactivecoventry.com

R. Iqbal (✉)
Coventry University, Coventry, UK
e-mail: r.iqbal@coventry.ac.uk

F. Doctor
Coventry University, Coventry CV1 5FB, UK
e-mail: f.doctor@coventry.ac.uk

I. Hatzilygeroudis and V. Palade (eds.), *Advances in Hybridization of Intelligent Methods*, Smart Innovation, Systems and Technologies 85,
https://doi.org/10.1007/978-3-319-66790-4_4

1 Introduction

The cost of traffic congestion is dramatically increasing both in terms of delays caused in commuting, delivery of goods and services and importantly CO_2 emissions. Big data analytics, IoT (Internet of Things) and machine learning approaches have the potential to address the traffic congestion problem more effectively by means of traffic modelling, visualisation and prediction for urban mobility management. These big-data based approaches are also fundamental in developing solutions for connected cars and autonomous vehicles. Big data analytics is used to examine and process high volume data to discover hidden patterns, reveal the underlying structure, identify relationships and gain other insight concerning traffic congestion and management. Recent advances in hardware and software technologies have facilitated big data acquisition. This data can now be harvested from a large number of diverse sources including GPS, IoT, traffic videos, phone utilisation and social media posts. IoT refers to the various sensors, actuators and controllers incorporated into Internet enabled devices to facilitate data exchange.

The utilisation of this huge amount of information has positioned big data and big data analytics as a centre of focus for research communities, businesses, and governments [8, 19]. These all work together to deliver new concepts, products and services which serve an ever-increasing number of diverse application contexts from smart cities [9] to healthcare [23]. This has contributed to an increase of the amount of data available and led to new challenges in the processing of the data. To overcome these problems a set of techniques have been developed all of which are grouped under the common name of machine learning (ML).

ML approaches are used for modelling patterns and correlations in data to discover relationships and make predictions, based on unseen data/events. ML approaches consist of supervised learning (learning from labelled data), unsupervised learning (discovering hidden patterns in data or extracting features) and reinforcement learning (goal oriented learning in dynamic situations). As such, ML approaches can also be categorised into: regression techniques, clustering approaches, density estimation methods and dimensionality reduction approaches. Non-exhaustive examples of these approaches are: decision tree learning, associate rule learning, artificial neural networks, deep learning support vector machines, clustering and Bayesian networks.

Computational Intelligence (CI) is a subcategory of ML approaches focusing on algorithms that are devised to imitate human information processing and reasoning mechanisms. These approaches are nature-inspired computational methodologies which have been developed to address complex real-world data-driven problems for which mathematical and traditional modelling are unable to work owing to: high complexity, uncertainty and stochastic nature of processes. Fuzzy Logic [1, 2, 6, 16], Evolutionary Algorithms [28] and Artificial Neural Networks [4] together comprise the core CI approaches that have been developed to handle this growing class of real-world problems.

This paper presents a part of our continuing research for big data and predictive analytics applying deep learning approaches to IoT and smart city applications. The rest of the paper is organised as follows. Section 2 discusses computational intelligence for big data analytics. Section 3 presents a novel methodology to provide solutions to data driven problems. Section 4 discusses the proposed hybrid based approach. Section 5 discusses a case study of the Taxi prediction application. Section 6 presents the performance of the Taxi prediction application. Finally, Sect. 7 presents conclusions and future directions.

2 Intelligent Transport System

The growth of modern urbanised and increasingly connected rural environments requires the development of efficient transportation infrastructures to better support the needs of visitors and commuters. Many initiatives have been implemented to satisfy personalised and contextualised user defined objectives to improve user mobility, utility, and satisfaction while helping to avoid congestion [5, 27].

The self learning car developed by Jaguar Land Rover is a an example of a state of the art vehicle exploiting some of these capabilities [14]. This emerging technological environment creates a massive opportunity for vehicles to take full advantage of the concept of the Internet of Things (IoT) by collecting, processing and aggregating this information and transforming it into knowledge required by their cars, other drivers, vehicles, and the wider society. For example having a distributed sensor network of cars being able to assess the density of traffic on the road can be used to optimize the traffic within a region, maximizing the road and public transport throughput.

For the advantages of Big Data and IoT to be exploited there is an increasing need for implementing intelligent systems and computational techniques which can reduce the complexity and cognitive burden on accessing and processing the large volumes of data generated in both embedded hardware and software based data analytics [1, 2, 11, 20]. Furthermore, personalised contents need to to be delivered through energy efficient wireless communication networks [24] in secure vehicular mobile cloud and vehicular ad hoc networks (VANET) [17, 18].

Another problem is managing transport networks effectively in urban areas [25]. For example, if the demand for taxi services is known, then a more efficient method for picking up passengers can be devised. This can be achieved by using historical and real-time data to predict the hot spots where taxis pick up and set down people in an urban area over the course of a day. By collecting data for the high and low predicted taxi demands over the urban area, contextual information pertaining to traffic conditions, geospatial distribution of the fleet and vehicle telematics, recommendations to taxi operators for the distribution and optimisation of taxi services can be provided.

Behaviour modelling could also be used to recommend real-time re-routings to satisfy personal objectives of road users while relieving traffic congestion. The problem of optimally managing the distribution of taxi services can also be tackled at large transportation hubs such as railway stations and airports to meet passenger demands. This could be achieved by applying vision processing algorithms to live CCTV camera feeds to create an airport taxi stand passenger queue tracking system. This system could then identify and count individuals waiting in a queue to estimate the number of people entering/exiting and queuing throughput over time. This information could then be used to measure length, growth rate and predict the wait time for each queue, which can be visualized in real-time and used to send notifications/alerts based on operator triggers and thresholds.

3 Smart City Projects

In this section a number of smart system projects being carried out in the various cities of the United Kingdom are summarised. Applications for these systems include: connected and self-driving cars; data visualisation for planning and management; pervasive sensing and IoT based solutions for road infrastructure monitoring; and predictive modelling for traffic congestion management. Table 1 lists several recent UK based smart-transport initiatives (UK Smart Cities Index, [26]).

Towards the development of smart city projects various researchers have used a range of techniques and data sources. For example, Restricted Boltzmann Machine (RBM) and Recurrent Neural Network (RNN) have been used to tackle tremendous

Table 1 UK smart transport systems/projects

	Project city	Smart city innovation
1	Milton keynes	Use of big data to build a smart transport system and battle congestion and pollution (Space in IOT, self-driving vehicle, public engagement)
2	Peterborough	Use of 4D virtual model as testbeds to test city-scale policy and planning implications as well as new technology impacts prior to adoption or deployment
3	Liverpool	Use of cost effective sensor technologies, data from city transport and local authorities, pedestrian movement, environmental data
4	Cambridge	Use of intelligent technology for big analytics for traffic management, air quality, energy network and health and social care
5	Ipswich, suffolk	Use of data to improve urban planning and transport systems, with further aspect of engaging local citizens via their mobile phones based on their precise physical location and context
6	East london, silvertown	Development of a connected society (application areas include from energy efficiency management and transport to waste management and parking, street furniture and way finding)
7	Bristol	Development of a physical and digital model of the city combined with power analytics and 3D visualisation tools
8	Reading and bracknell	Development of IoT based solutions to provide smart transport solutions for highways and public transport services

high-dimensional traffic data from taxi to model and predict traffic congestion [22]. Another project has used Artificial Neural Networks for the real-time prediction of bus travel speeds [15]. Object detection and identification for traffic monitoring was carried out using Convolutional Neural Network [3]. The algorithm was applied on traffic monitoring videos. Several researchers have used social media for the prediction of traffic congestion [7].

Table 2 Data sources and type of data

	Data source	Type of data
1	Local/government authorities	Local/government internal data
2	Global positioning system (GPS)	Data from taxi/bus companies Companies that provide navigation and mapping products
3	Monitoring equipment	Traffic videos Thermal imaging cameras
4	Telecommunication providers	Phone utilisation
5	Web based sources	Social media posts RSS feeds to identify significant events
6	Academic experts and research literature	Subject specific knowledge

Table 3 Sensors for traffic control and monitoring systems

	Sensors /devices for traffic control and monitoring	Description
1	Loop detectors	To detect vehicle and measure velocity as well as transmit the collect data
2	Radar	To capture traffic data (velocity, occupancy and flow rate of vehicles)
3	Video	For traffic monitoring and to record vehicle number plates
4	License Plate Readers	To capture number plate information. Other traffic information can also be captured by the use of multiple devices
5	Radio-frequency identification	Collecting tolls from drivers, measuring travel times
6	Bluetooth	To calculate travel time when multiple readers are used on roads
7	Wireless sensors	To provide similar functionality as loop detectors. These devices will be able to provide more accurate information
8	Sparsely sampled GPS	Provides information about the location of vehicles. The data is collected from GPS enabled vehicles at a fixed frequency
9	High frequency GPS	Provides more accurate information about the location of vehicles

Smart city projects have used various sources of data to capture information related to the location and speed of the traffic. The various data sources and the type of data which can be collected including equipment and devices, as shown in Table 2 are briefly explained. This section also discusses various other sensors or dedicated devices used for data acquisition as shown in Table 3.

4 Proposed Hybrid Approach

In this section, a hybrid approach that can be applied for the analysis and modelling of smart city data sources is presented. This approach has already been used successfully in the development of the taxi demand prediction application described in Sect. 5. It provides solutions to other challenging real world big data problems that require spatial-temporal modelling.

The core part of this approach is based on the latest discoveries in the field of neuroscience and introduces a novel universal generative modelling approach called Hierarchical Spatial-Temporal State Machine (HSTSM). The model is inspired and implemented based on the understanding of the structure and functionality of the human brain. The implementation is based on a hybrid method which includes deep belief networks, auto-encoders, agglomerative hierarchical clustering and temporal sequence processing [21]. The model handles high volumes of data characterised by complex correlations between the input values and temporal consequences of the different input states of the system (termed here as spatial-temporal correlations).

The main elements of the proposed methodology include the data layer, input layer, data transformation layer, spatial pooling, temporal inferences, prediction model and finally, the presentation of information or application layer as shown in Fig. 1.

The data layer captures and aggregates heterogeneous data from many data sources including hardware and IoT, manual input and software.

The Data Transformation Layer encodes the input data into binary distributed representations, scalable to high dimensional data. Spatial Modelling identifies a hierarchical organisation of multiple levels of data abstraction achieved in a process of automatic feature extraction.

The discovery of correlations between individual inputs are determined by the spatial transformation (spatial pooling) of input space into a feature space and the discovery of correlations between the individual inputs [10]. To achieve this, the encoded vector is compressed and an automatic process of feature extraction is performed by deep belief neural networks (DBN). The trained restricted Boltzmann Machines (RBM) forming the DBN are used to initialise the deep auto-encoder. This unsupervised method of feature extraction enables the model to acquire an improved and more compressed representation of the input space. Hierarchical clustering is performed on the transformed features derived from the deep belief network, to extract the many possible states of the modelled system. The main purpose of this operation is to reduce the input space to a fixed number of the most

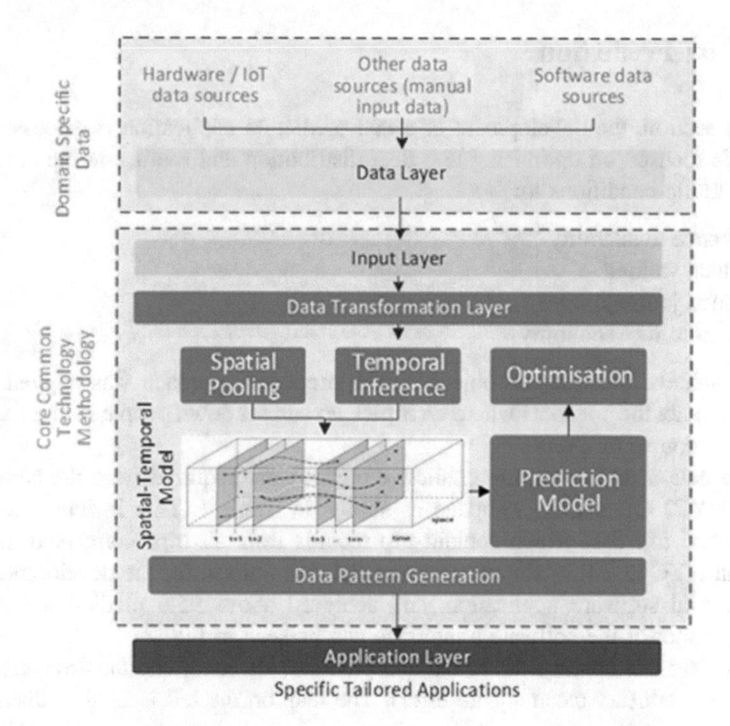

Fig. 1 Proposed hybrid approach for big data analytics

probable states. The basic metric used with the hierarchical clustering is Euclidian distance. This process is a type of spatial pooling, where the original binary inputs that occur close together in space are pooled together.

Temporal inference is performed on the identified states, and prediction of the next possible state of the system can be achieved with the utilisation of an n-order Markov chain. Prediction acquired in this way can be subsequently used to identify specific patterns of behaviour of the modelling problem under investigation. At the end of this process, the predicted vector can be compared with the actual input vector, generated at each moment in a specific application context, to identify specific patterns or irregular behaviour. This can be achieved by distance function or basic ML techniques, like a Multilayer Perceptron. Spatial-temporal model achieved in this way can also serve as an input to various optimisation frameworks to optimise the processes being modelled.

The application layer is responsible for displaying the results through the interface of an application or visualisation which can provide stakeholder insight into the data being modelled. The visualisation or interface can also be personalised using the type and functional form of information representation to meet the individual needs of users [12, 13].

5 Taxi Prediction

In this section, the development of a taxi prediction application is discussed. The work is focused on optimising taxi fleet distribution and routing in the context of urban traffic conditions to:

- enhance availability
- reduce waiting
- reduce journey times
- promote fuel economy

To achieve the research objectives the proposed approach was applied which could predict the hot spots where taxis pick up and set down people in an urban area over the course of a day.

The data used to train the predictive model were acquired from the New York City (NYC) Open Data webpage as shown in Table 4. This includes the 2013 Green taxi trip data which contain trip records from all trips completed in green taxis in NYC in 2013. The proposed model was utilised for the development of a commercial software application, and achieved above 95% prediction accuracy. A screenshot of the software's interface can be seen in Fig. 2.

The user can observe the actual and predicted demand, and the error generated by the system (i.e. mean square error). The map on the left is used to display the

Table 4 NYC dataset features

Field	Value
Pickup_datetime	12/06/2013 12:11:05 PM
Dropoff_datetime	12/06/2013 12:14:06 PM
Store_and_fwd_flag	N
Rate_code	1
Dropoff_latitude	40.811981201171875
Passenger_count	1
Trip_distance	0.5
Fare_amount	4
Extra	0
MTA_tax	0.5
Tip_amount	0
Tolls_amount	0
Ehail_fee	
Total_amount	4.5
Payment_type	2
Trip_type	
Pickup_longitude	−73.9650650024414
Pickup_latitude	40.8061408996582
Dropoff_longitude	−73.96223449707031
Vendor_id	1

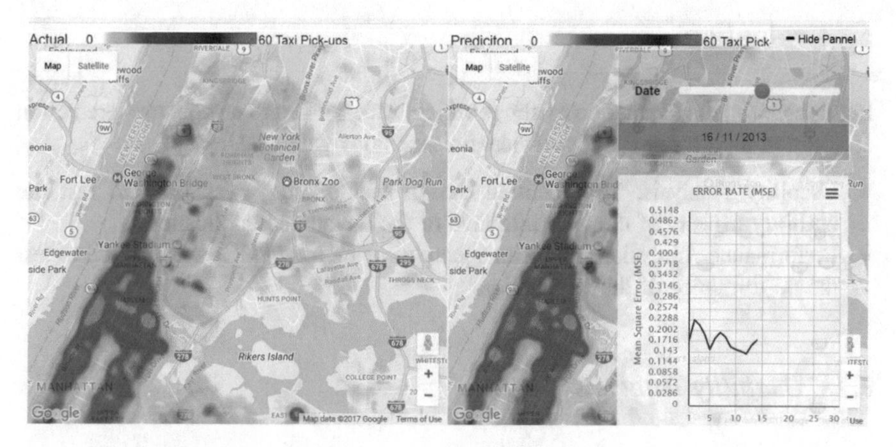

Fig. 2 Taxi demand prediction application interface (Iteration 4)

actual taxi pick-ups while the map on the right shows the predicted taxi pick-ups. A "–/+" button is used to hide or show the main control panel. The Control Panel (top left side) has an adjustable slider bar for selecting the prediction date with a label indicating the selected date.

The application shows and predicts hot spots where taxi's drop off and set down people in urban areas over the course of a day. This information can be fed to an optimisation system using evolutionary algorithms to optimise the real-time distribution of taxi's based on the predicted demand (i.e. from the developed prediction model) together with consideration of other factors pertaining to fleet load, traffic situation and optimal routing, urban events, status of the fleet, running costs and possibly environmental considerations such as weather and the carbon footprint of the fleet operations. The system could then recommend customers taxi availability/advisability, estimation of routings, waiting times and journey times based on their location, what they are doing and the period of engagement. For instance: dinner in restaurant 'A' in location 'Y' reserved for 7 pm for approx. 1.5 h, followed by a show in location 'X' starting at 9 pm for 2 h on day Z, where factor 1 is 'B' and factor n is 'C'.

Based on the above scenario, a pre-emptive recommendation of taxi services availability, waiting times and journey times can be made to users via their smart phones. Alternatively, an auto booking could be made based on the predicted need given the personal information about user activities, schedule of appointments, engagement and transport utilisation behaviour.

6 Evaluation

In this section, the evaluation methods which were applied to test the taxi prediction application are discussed. Two evaluation methods were used.

- Usability Analysis: User centred design methodology was used to test the usability aspects of the web interface of the developed application.
- Performance Analysis: The mean square error metric (MSE) was applied to test the performance of the application.

6.1 Usability Analysis

The usability of the web based application has been tested using the user centred methodology as discussed in [12, 13]. The usability analysis was carried out by 7 participants involving 2 members of the design team, 1 usability expert and 4 end users. An iterative process was followed which led to the development of four versions of the application. The main features of each iteration are briefly explained.

First Iteration

One web based interface was designed to display either the predicted or actual data at any given time by pressing the button labelled as Prediction or Actual.

Second Iteration

This iteration uses a split screen to show predicted or actual data on one screen (similar to the one shown in Fig. 2 but without displaying the mean square error rate chat).

Third Iteration

An interface was designed to show the predicted and actual data on one screen as hot spots with different colours. In this iteration, it was difficult for the end users to view and compare predictive and actual data points (See Fig. 3).

Fourth Iteration

One screen showing both predicted and actual points in one screen with the mean square error rate chart. The user interface (iteration 4) which met user requirements is shown in Fig. 2.

6.2 Performance Analysis

The data used to train the predictive model has been acquired from the NYC Open Data webpage. The 2013 Green taxi trip data includes trip records from all trips by

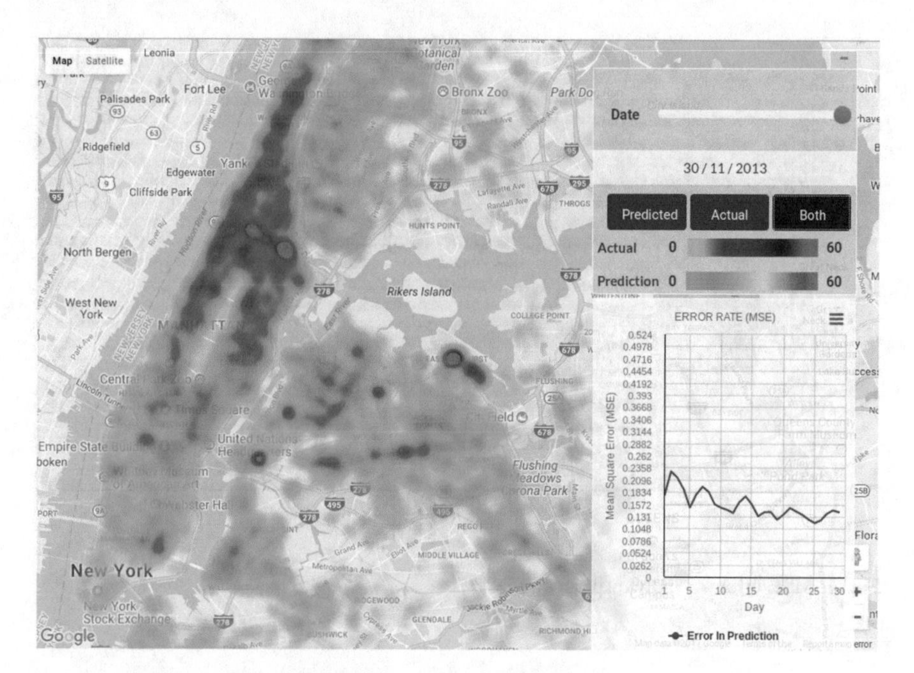

Fig. 3 Taxi demand prediction application interface (Iteration 3)

NYC green taxis in 2013. Records include fields capturing pick-up and drop-off dates/times, pick-up and drop-off locations, trip distances, itemised fares, rate types, payment types, and driver-reported passenger counts. The data used in the attached datasets were collected and provided to the NYC Taxi and Limousine Commission (TLC) by technology providers authorised under the Livery Passenger Enhancement Program (LPEP). An exemplary row in the data set is shown in Table 4.

The dataset used in this application was ranged for 11 months from January 2012 to November 2013. In order to train and evaluate our predictive model, we split the data set into two folders as shown in Table 5. One month data (December 2013) was not used owing to additional noise in the data caused by holidays season.

To assess the performance of the application the mean square error metric (MSE) was used. MSE is a measure of performance which is widely used in machine learning and statistical application to determine the error in the predictive model against test data. The equation is shown below.

Table 5 Training and test data set split

No	Data	Ratio%	Description
1	Training data	90.9	10 months data from January 2013-Oct 2013
2	Test data	9.1	1 month data for November 2013

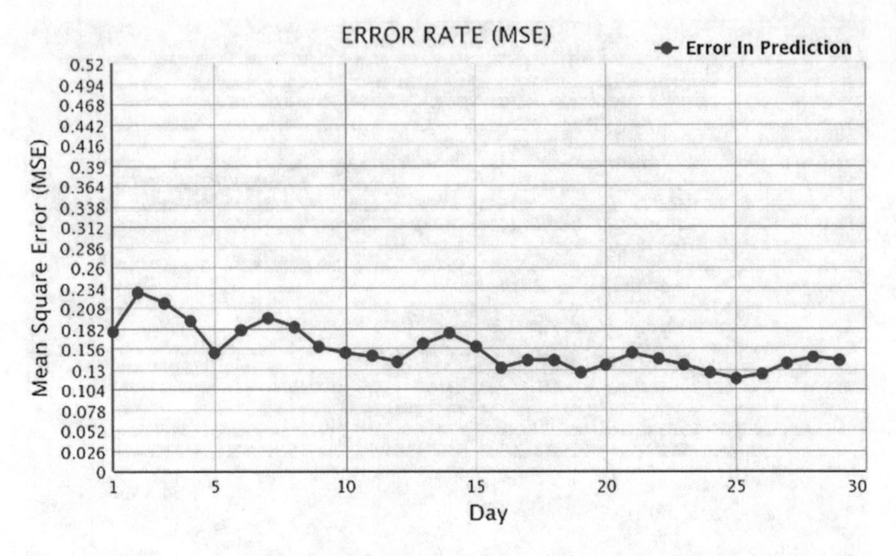

Fig. 4 MSE error for individual days in test set

$$MSE = \frac{1}{N}\sum_{i=1}^{n}(Z_i - Y_i)^2$$

where Z_i is predicted value, Y_i is a target value and N is a number of samples.

The metric has been embedded into the application interface which displays the error rate for current prediction (per day) corresponding to the slider position. The accuracy of the predictions was above 95% based on the unseen data. The chart presents the mean square error (MSE) of the predictions for a given day. The values on Y axis give the average error of predictions for each point on the chart. For example, 0.3 meant that on average if each point on the map is considered, the predictions could be wrong by 0.3 Taxi pickups. Each Taxi pick up location on the map had an accuracy of up to 5 m, meaning the actual pick up location point could be within 5 m from the predicted location. The X axis shows the error for a specific day. Figure 4 shows the individual MSE for test dates between 1st November 2013 and 30th November 2013.

7 Conclusions

In this paper, the importance of big data analytics and the application of machine learning in urban mobility and smart city applications has been discussed. Special attention has been given to modelling and predicting traffic from GPS and temporal data and an overview of computational intelligence techniques appropriate for the effective processing and analysis of big data has also been discussed.

A data modelling methodology, which introduces a novel biologically inspired universal generative modelling approach called Hierarchical Spatial-Temporal State Machine (HSTSM) has been presented. The proposed methodology relies on a hybrid method, based on the structure and functions of the mammalian brain. It incorporates different soft computing techniques and has the potential to deal with large amounts of data, which are characterised by spatial-temporal correlations. An application case study has been provided that demonstrates the use of a novel deep learning spatial modelling technique. The method is applied to the prediction of hot spots where taxis pick up and set down people in an urban area over the course of a day.

The potential benefits arising from this novel methodology are numerous spanning a large spectrum of smart city application areas such as traffic congestion prediction, anticipating maintenance and overhaul requirements of public transport, fleet management and optimising deployment based utilisation patterns and consumer data driven improvements in urban services. These solutions can have a significant impact on knowledge, society, the economy, and individuals. Scientific knowledge and research may benefit from revealing hidden patterns in data providing city and the insight of municipal authorities to drive urban planning and management policies. Society could profit from the delivery of applications, which promote improved public transportation services. E-businesses and organisations could also be assisted through sentiment analysis tools, which could contribute to the delivery of personalised and contextualised products and services, meeting the needs of urban users, commuters and visitors.

Future work will involve the utilisation and deployment of the proposed methodology to different application areas to create novel models and applications with significant commercial and scientific value together with further improvements of the developed systems.

References

1. Akuma, S., Iqbal, R., Jayne, C., Doctor, F.: Comparative analysis of relevance feedback methods based on two user studies. Comput. Hum. Behav. **60**, pp. 138–146, Elsevier (2016)
2. Alhabashneh, O., Iqbal, R., Doctor, F., James, A.: Fuzzy rule based profiling approach for enterprise information seeking and retrieval. Inf. Sci. (2017)
3. Bautista, C.M., Dy, C.A., Mañalac, M.I., Orbe, R.A., Cordel, M.: Convolutional neural network for vehicle detection in low resolution traffic videos. In: IEEE Region 10 Symposium (TENSYMP), pp. 277–281. IEEE (2016)
4. Campo, I., Finker, R., Martinez, M., Echanobe, J., Doctor, F.: A real-time driver identification system based on artificial neural networks and cepstral analysis. In: International Joint Conference on Neural Networks (IJCNN), pp. 1848–1855. IEEE (2014)
5. Djahel, S., Jabeur, N., Barrett, R., Murphy, J.: Toward V2I communication technology-based solution for reducing road traffic congestion in smart cities. In: International Symposium on Networks, Computers and Communications (ISNCC), pp. 1–6. IEEE (2015)

6. Doctor, F., Syue, C.H., Liu, Y.X., Shieh, J.S., Iqbal, R.: Type-2 fuzzy sets applied to multivariable self-organizing fuzzy logic controllers for regulating anesthesia. Appl. Soft Comput. **38**, 872–889 (2016)
7. Gong, Y., Deng, F., Sinnott, R.O.: Identification of (near) Real-time traffic congestion in the cities of australia through twitter. In: Proceedings of the ACM First International Workshop on Understanding the City with Urban Informaticspp, pp. 7–12. ACM (2015)
8. Hashem, I.A.T., Yaqoob, I., Anuar, N.B., Mokhtar, S., Gani, A., Khan, S.U.: The rise of "big data" on cloud computing: Review and open research issues. Inf. Syst. **47**, 98–115 (2015)
9. Hashem, I.A.T., Chang, V., Anuar, N.B., Adewole, K., Yaqoob, I., Gani, A., Chiroma, H.: The role of big data in smart city. Int. J. Inf. Manag. **36**(5), 748–758 (2016)
10. Hawkins, J.: On Intelligence. Times Books, New York (2004)
11. Iqbal, R., Grzywaczewski, A., Halloran, J., Doctor F., Iqbal, K.: Design implications for task-specific search utilities for retrieval and reengineering of code, Enterprise Information Systems, Taylor and Francis, pp. 1751–7575 (2015) doi:10.1080/17517575.2015.1086494
12. Iqbal, R., Shah, N., James, A., Duursma, J.: ARREST from work practices to redesign for usability. Int. J. Expert Syst. Appl. **38**(2), 1182–1192 Elsevier (2011)
13. Iqbal, R., Sturm, J., Kulyk, O., Wang, C., Terken, J.: User-centred design and evaluation of ubiquitous services. In: Proceedings of the 23rd Annual International Conference on Design of Communication: Documenting and Designing for Pervasive Information, ACM SIGDOC, pp. 138–145. ISBN: 1-59593-175-9 (2005)
14. Jaguar Land Rover Limited Predicting a current destination from a combination of user data and activity data. Patent reference number: 1412167.7 (2014)
15. Julio, N., Giesen, R., Lizana, P.: Real-time prediction of bus travel speeds using traffic shockwaves and machine learning algorithms. Res. Transp. Econ. (2016)
16. Karyotis, C., Doctor, F., Iqbal, R., James, A.: A fuzzy computational model of emotion for cloud based sentiment analysis. Inf. Sci. (2017)
17. Kumar, N., Iqbal, R., Mistra, S., Rodrigues, J.: Bayesian coalition game for contention aware reliable data forwarding in vehicular mobile cloud. J. Future Gener. Comput. Syst. **48**, 60–72. Elsevier (2014)
18. Kumar, N., Iqbal, R., Misra, S., Rodrigues, J.J.P.C : Bayesian cooperative coalition game as-a-service for RFID-based secure QoS management in mobile cloud. IEEE Trans. Emerg. Top. Comput. **99**, IEEE Press (2016)
19. Mahmud, S., Iqbal, R., Doctor, F.: Cloud enabled data analytics and visualization framework for health-shocks prediction. J. Future Gener. Comput. Syst. **65**, 169–181. Elsevier (2016)
20. Maniak, T., Jayne, C., Iqbal, R., Doctor, F.: Automated intelligent system for sound signalling devicequality assurance. Inf. Sci. **294**, 600–611. Elsevier (2015)
21. Maniak, T., Iqbal, R., Doctor, F.: A Method for Monitoring the Operational State of a System. Interactive Coventry Ltd Pending Patent Reference No: 1607820.6 (2016)
22. Ma, X., Yu, H., Wang, Y., Wang, Y.: Large-scale transportation network congestion evolution prediction using deep learning theory. PLoS ONE **10**(3), e0119044 (2015)
23. Murdoch, T.B., Detsky, A.S.: The inevitable application of big data to health care. JAMA **309** (13), 1351–1352 (2013)
24. Qureshi, F.F., Iqbal, R., Asghar, M.N.: Energy efficient wireless communication technique based on cognitive radio for internet of things. J. Netw. Comput. Appl. (2017)
25. Tirachini, A.: Estimation of travel time and the benefits of upgrading the fare payment technology in urban bus services. Transp. Res. Part C: Emerg. Technol. **30**, 239–256 (2013)
26. UK Smart Cities Index. Assessment of Strategy and Execution of the UK's Leading Smart Cities, Commissioned by Huawei, Navigant Consulting, Inc.17 May 2016
27. Vegni, A.M., Biagi, M., Cusani, R.: Smart Vehicles, Technologies and Main Applications in Vehicular ad hoc Networks. INTECH, Open Access Publisher (2013)
28. Whitley, D.: An overview of evolutionary algorithms: Practical issues and common pitfalls. Inf. Softw. Technol. **43**(14), 817–831 (2001)

Assurance in Reinforcement Learning Using Quantitative Verification

George Mason, Radu Calinescu, Daniel Kudenko and Alec Banks

Abstract Reinforcement learning (RL) agents converge to optimal solutions for sequential decision making problems. Although increasingly successful, RL cannot be used in applications where unpredictable agent behaviour may have significant unintended negative consequences. We address this limitation by introducing an *assured reinforcement learning* (ARL) method which uses quantitative verification (QV) to restrict the agent behaviour to areas that satisfy safety, reliability and performance constraints specified in probabilistic temporal logic. To this end, ARL builds an abstract Markov decision process (AMDP) that models the problem to solve at a high level, and uses QV to identify a set of Pareto-optimal AMDP policies that satisfy the constraints. These formally verified *abstract policies* define areas of the agent behaviour space where RL can occur without constraint violations. We show the effectiveness of our ARL method through two case studies: a benchmark flag-collection navigation task and an assisted-living planning system.

Keywords Reinforcement learning · Safety · Quantitative verification · Abstract markov decision processes

---

G. Mason (✉) · R. Calinescu · D. Kudenko
Department of Computer Science, University of York, Deramore Lane, York, UK
e-mail: grm504@york.ac.uk

R. Calinescu
e-mail: radu.calinescu@york.ac.uk

D. Kudenko
e-mail: daniel.kudenko@york.ac.uk

D. Kudenko
Saint Petersburg National Research Academic University of the Russian Academy of Sciences, St. Petersburg, Russia

A. Banks
Defence Science and Technology Laboratory, Salisbury, UK
e-mail: abanks@mail.dstl.gov.uk

I. Hatzilygeroudis and V. Palade (eds.), *Advances in Hybridization of Intelligent Methods*, Smart Innovation, Systems and Technologies 85,
https://doi.org/10.1007/978-3-319-66790-4_5

1 Introduction

Reinforcement learning (RL) [1] is a widely-used machine learning technique used to find solutions for sequential decision making problems. Solutions are learned by an autonomous agent which explores an initially unknown Markov decision process (MDP) to find an optimal *policy*, i.e. the actions to take in different MDP states in order to maximise the cumulative reward while navigating the MDP. Despite successful adoption in some robotics [2], sensing [3], gaming [4] and control [5] applications, traditional RL cannot be used in mission- and safety-critical applications. In these applications, unpredictable agent actions can lead to mission failure, increased risks to humans or other systems, or violations of legal requirements (e.g. in business domains) [6]. Furthermore, if the system learns a solution that does not conform to conventional behaviours expected of its domain, the users of the system may have difficulty trusting the solution even though it may not violate any requirements.

The difficulty lies in how objectives are defined in RL and how solutions are learned by the agent [7]. Objectives are imparted in an RL MDP by a *reward structure* which returns numerical rewards to the agent for achieving objectives. The agent aims to maximizing the reward cumulated and therefore achieve as many objectives as possible. However, in scenarios where objectives conflict with one another it becomes difficult to define an effective reward structure. For example, in a case study we describe later in this paper, the agent's objective is to traverse an area to reach as many checkpoints as possible (yielding a positive reward) whilst minimizing the risk of being captured (which yields a negative reward, i.e. a punishment). The agent can either attempt to reach all the checkpoints at high risk of capture, or it can ignore them all with no risk of capture. Defining a maximum probability of risk that the agent is captured and minimum number of checkpoints it must reach is not possible with traditional RL, which cannot accommodate finding solutions that lie within a specific range of probability and a specific range of rewards. Traditional RL can maximize (or minimize) one objective but at the detriment of the other. Alternatively, it can learn to optimize a solution to compromise for both objectives at the same time, but the solution is unlikely to satisfy strict safety requirements. Furthermore, it is often the case that complex requirements can be difficult to define concisely in the reward function, requiring more features to define environment states, expanding the state space, thereby worsening the state space explosion problem and significantly increasing the time taken to learn solutions [1].

Recently there has been increasing research into addressing this problem [7]. Approaches range from changing how reward accumulation is optimised to how the agent actively explores the environment. However, these approaches are difficult to generalize, cannot provide firm guarantees that the solution will be safe or, where safety can be assured, it is at a significant cost to the optimality of the solution and difficulty in expressing realistic safety properties.

In this extended version of our preliminary work on providing assurance for RL policies [8], we present and extensively evaluate an RL method which can be used in applications that must satisfy strict safety, reliability and performance constraints. Our *assured reinforcement learning* (ARL) combines traditional reinforcement learning with a formal analysis stage in which the agent exploration is restricted to areas of the original MDP that satisfy the required constraints. ARL carries out this analysis using *quantitative verification* (QV) [9], a mathematically based technique for establishing the reliability, performance and other quality-of-service properties of stochastic systems. Particular advantages of our hybrid ARL are scalability due to a hierarchical approach and convenience of formulating required constraints by using an expressive representation language that has been successfully applied in quantitative verification.

Specifically, ARL supports constraints specified in a variant of probabilistic temporal logic called *probabilistic computation tree logic* (PCTL) [10], and comprises a QV stage and an RL stage. In the QV stage, expert-provided knowledge of the scenario is given in the form of an abstract Markov decision process (AMDP) [11], a common and feasible practice in safety engineering [12–14]. Compared to the complete MDP to be explored by the RL agent, the AMDP can be assembled with only limited understanding of the problem, and has a significantly reduced state space and a simplified action set [11, 15, 16] that enable efficient analysis which would not have been feasible without this hierarchical approach. Given a set of PCTL constraints, quantitative verification is then used to identify AMDP policies that satisfy all these constraints. Each of these "safe" *abstract policies* resolves some of the nondeterminism of the original MDP, inducing a restricted MDP that the agent explores in the reinforcement learning stage of our ARL method without violating any of the constraints.

As described above, ARL incorporates a set of constraints on the behaviour of a reinforcement learning agent both in the learning process and in the learnt policy. Multiple "safe" abstract policies are typically generated during the QV stage. ARL supports the selection of a suitable abstract policy for the RL stage by retaining only the abstract policies that are *Pareto-optimal* with respect to optimization objectives associated with constraints from the QV stage and/or specified additionally.

Our work contributes to the ongoing research on *safe reinforcement learning* [7]. Thus, ARL complements the existing *constrained optimisation* approaches to safe RL, in which the agent seeks a policy that maximises its obtained reward subject to a set of constraints. To the best of our knowledge, ARL is the first such approach that supports the broad range of safety, reliability and performance constraints that can be formally specified in PCTL [10] extended with rewards [17], and that uses quantitative verification [9] to identify *allowable MDP policies*. In contrast, the existing approaches are typically limited to specifying bounds for the reward obtained by the RL agent or for simple measures related to its optimisation [18–23].

The remainder of this paper is organized as follows. Section 2 introduces the technologies that are used in ARL. Section 3 provides an example scenario, based on the benchmark RL flag-collection domain [24] and modified to include an aspect of risk where the application of ARL is necessary. Section 4 outlines the procedure for using ARL, using the example scenario to illustrate the process. Section 5 evaluates ARL through two case studies, the first based on the running example and the second based on an assisted-living system [25]. Section 6 discusses related research, and Sect. 7 summarizes our results and discusses directions for future work.

2 Background

2.1 Markov Decision Processes (MDPs)

Markov decision processes represent a formalism for modelling sequential decision-making problems [1]. An MDP models an environment in which an agent (i.e. decision maker) can perceive the current state s, and select an action a from a set of actions. States contains *features* which uniquely define the status of the environment. Performing the selected action a results in the agent transitioning to a new state s' and receiving an immediate numerical reward $r \in \mathbb{R}$.

It is through rewards that objectives are defined in an MDP. Generally, objectives that the agent must aim to achieve are given a positive reward and events and areas in the MDP that the agent must avoid (e.g. fail states) are assigned a negative reward (also referred to as a *cost*). The agent aims to maximize the rewards it cumulates, thus achieving objectives, or inversely, minimize the cost of its actions.

Formally, an MDP is defined as a tuple (S, A, T, R), where:

- S is a set of states;
- A is a set of actions;
- T is a state transition function such that for any $s, s' \in S$ and any action $a \in A$ that is permitted in state s, $T(s, a, s')$ gives the probability of transitioning to state s' when performing action a in state s;
- R is a reward function such that $R(s, a, s') = r$ is the reward received by the agent when action a performed in state s leads to state s'.

A related concept that is central to RL is that of a *policy*. ARL uses *deterministic policies*, i.e. mappings of the form $\pi : S \rightarrow A$ that associate each state $s \in S$ to one of the actions allowed in state s.

When all elements of the MDP are known, the problem can be solved using dynamic programming, e.g. by using the value or policy iteration algorithms. In scenarios where the transition and/or reward functions are unknown a priori, RL needs to be used as described in Sect. 2.3.

2.2 Quantitative Verification (QV)

QV is a formal verification technique used to establish safety, reliability, performance and other non-functional properties of systems through the analysis of their stochastic models [9, 26]. Unlike techniques like testing and simulation, QV uses efficient algorithms to examine the entire state space of the analysed model, yielding results that are guaranteed to be correct. QV supports the analysis of models including MDPs, Markov chains and probabilistic automata. The analysed properties of these models are specified formally in probabilistic variants of temporal logic. QV is performed using efficient probabilistic model checkers, such as PRISM [27] or MRMC [28].

For the analysis of MDPs, QV labels the model states with *atomic propositions* that specify basic properties of interest that hold in each MDP state, e.g. *success*, *fail* or *retry*. MDPs labelled with atomic propositions enable the QV of properties that express probabilities and temporal relationships between events. For example, QV can verify if the probability of achieving *success* without any *retry* (i.e. of reaching a state labelled *success* without passing through a state labelled *retry*) is at least 0.95. These properties are specified in a probabilistic temporal logic called probabilistic computational tree logic (PCTL) [10]. Given a set of atomic propositions AP, a *state formula* Φ and a *path formula* Ψ in PCTL are defined by the grammar:

$$\begin{array}{l} \Phi ::= true \mid a \mid \neg\Phi \mid \Phi_1 \wedge \Phi_2 \mid \mathrm{P}_{\bowtie p}[\Psi] \\ \Psi ::= \mathrm{X}\Phi \mid \Phi_1 \,\mathrm{U}\, \Phi_2 \mid \Phi_1 \,\mathrm{U}^{\leq k}\, \Phi_2 \end{array}, \tag{1}$$

where $a \in AP$, $\bowtie \in \{<, \leq, \geq, >\}$, $p \in [0, 1]$ and $k \in \mathbb{N}$; and a PCTL *reward state formula* [29] is defined by the grammar:

$$\Phi ::= \mathrm{R}_{\bowtie r}[\mathrm{I}^{=k}] \mid \mathrm{R}_{\bowtie r}[\mathrm{C}^{\leq k}] \mid \mathrm{R}_{\bowtie r}[\mathrm{F}\Phi] \mid \mathrm{R}_{\bowtie r}[\mathrm{S}], \tag{2}$$

where $r \in \mathbb{R}_{\geq 0}$. State formulae include the logical operators $\wedge$ and $\neg$, which allow the formulation of disjunction ($\vee$) and implication ($\Rightarrow$).

The semantics of PCTL are defined with a satisfaction relation $\models$ over the states and paths of an MDP (S, A, T, R). Thus, $s \models \Phi$ means Φ is satisfied in state s. For any state $s \in S$, we have: $s \models true$; $s \models a$ iff s is labelled with the atomic proposition a; $s \models \neg\Phi$ iff $\neg(s \models \Phi)$; and $s \models \Phi_1 \wedge \Phi_2$ iff $s \models \Phi_1$ and $s \models \Phi_2$. A state formula $\mathcal{P}_{\bowtie p}[\Psi]$ is satisfied in a state s if the probability of the future evolution of the system satisfying Ψ satisfies $\bowtie p$. For an MDP path $s_1 s_2 s_3 \ldots$, the "next state" formula $\mathrm{X}\,\Phi$ holds iff Φ is satisfied in the next path state (i.e. in state s_2); the *bounded until* formula $\Phi_1\, \mathrm{U}^{\leq k}\, \Phi_2$ holds iff before Φ_2 becomes true is some state s_x, $x < k$, Φ_1 is satisfied for states s_1 to s_{x-1}; and the *unbounded until* formula $\Phi_1\, \mathrm{U}\, \Phi_2$ removes the constraint $x < k$ from the bounded until. For instance, the PCTL formula $\mathrm{P}_{\geq 0.95}[\neg retry\, \mathrm{U}\, success]$ formalises the constraint 'the probability of reaching *success* without *retry* is at least 0.95' from the earlier example.

The notation $\mathrm{F}^{\leq k}\Phi \equiv true\mathrm{U}^{\leq k}\Phi$, and $\mathrm{F}\Phi \equiv true\mathrm{U}\Phi$ is used when the first part of a bounded until, and until formula, respectively, are *true*. The reward state formulae (2) express the expected cost at timestep k ($\mathcal{R}_{\bowtie r}[\mathrm{I}^{=k}]$), the expected cumulative cost up to time step k ($\mathcal{R}_{\bowtie r}[\mathrm{C}^{\leq k}]$), the expected cumulative cost to reach a future state that satisfies a property Φ ($\mathcal{R}_{\bowtie r}[\mathrm{F}\Phi]$), and the expected steady-state reward in the long run ($\mathcal{R}_{\bowtie r}[\mathrm{S}]$).

Finally, probabilistic model checkers also support PCTL formulae in which the bounds '$\bowtie p$' and '$\bowtie r$' are replaced with '=?', to indicate that the computation of the actual bound is required. For example, $P_{=?}[F^{\leq 20}\, success]$ expresses the probability of succeeding (i.e. of reaching a state labelled *success*) within 20 time steps.

2.3 Reinforcement Learning (RL)

RL is used to solve MDPs when either the reward and/or transition function is unknown, and involves the use of an autonomous agent. The agent starts with no knowledge of the environment, and must learn about it by *exploration*, i.e. by selecting initially arbitrary actions whilst moving from one state of the unknown MDP to another.

By receiving rewards after each state transition, the agent learns about the quality of its action choices. The agent stores this knowledge it gains about the quality of a state-action pair (s, a) in the form of a *Q-value*, denoted $Q(s, a)$. Updates to Q-values are done iteratively using a temporal difference learning algorithm, such as Q-learning [30], and through these updates information about rewards in the environment are propagated over state-action pairs. The Q-learning algorithm has the update formula:

$$Q(s,a) \leftarrow Q(s,a) + \alpha[r + \gamma \cdot \max_{a' \in A} Q(s', a') - Q(s,a)], \tag{3}$$

where $\alpha \in (0, 1]$ is the learning rate and $\gamma \in [0, 1]$ is the discount factor. The learning rate determines how influential the rewards received are when updating Q-values. If the learning rate is too high then the agent may fail to converge on an optimal solution as the agent can oscillate around an optimal solution but fail to land on it. For this reason it is sometimes necessary to decay α to zero over the learning process. The discount factor specifies how far into the future the agent should consider, e.g. if the agent should perform an action which in the immediate future might incur a cost but in the distant future may yield a high reward.

With these Q-values the agent can exploit the knowledge it has already learned: when revisiting a state, instead of randomly selecting an action to perform, it can select an action based on a pre-defined policy. An example of such a policy is the *ϵ-greedy policy*, where with probability $0 < \epsilon < 1$ the agent will act randomly and with probability $1-\epsilon$ it will select the highest-value action it knows about [1]. Setting a value of ϵ too high can cause the agent to ignore previously learned information making it repeat actions that it has already learned may be costly, increasing the time it takes to learn an optimal solution. Similarly, though, if ϵ is too small then the agent will not explore enough, instead relying too heavily on the limited knowledge it has previously acquired, also prolonging the time taken to reach a solution.

An RL experiment comprises a number of learning *episodes*, collectively known as a learning *run*. In an episode, the agent performs actions until either it encounters an absorbing state or a maximum number of timesteps has passed, in both cases the learning episode will terminate. It is rarely sufficient to perform just a single episode, though, and multiple learning episodes are often required. For each successive episode the agent begins again from the starting state but retains the knowledge it has learned in previous episodes.

Provided that each state in the MDP is visited an infinite number of times the algorithm is guaranteed to converge to an optimal policy, i.e. the Q-values no longer change when updated. In practice, though, a finite number of learning episodes is typically sufficient to obtain an optimal policy, or one that is sufficiently optimal.

2.4 Abstract MDPs (AMDPs)

An AMDP is a high-level representation of an MDP in which multiple states of the MDP are aggregated (e.g. based on their similarity [15]). Additionally, the low-level actions of the MDP are replaced by temporally abstract *options* [16]. For example, instead of an agent performing a sequence of stepwise movements to transition through a series of Cartesian coordinates from location A to enter location B, in an AMDP each location would be a single state and the option would simply be to "move" from one location to the other. An AMDP is orders of magnitude smaller than its MDP counterpart, can often be assembled with significantly less knowledge about the environment, and can be solved and reasoned about much faster [11].

Given an MDP (S, A, T, R) and a function $z : S \rightarrow \bar{S}$ that maps each state $s \in S$ to an abstract state $z(s) \in \bar{S}$ such that $z(S) = \bar{S}$, an AMDP can be formally defined as a tuple $(\bar{S}, \bar{A}, \bar{T}, \bar{R})$, where:

- $\bar{S}$ is the set of abstract states;
- $\bar{A}$ is the set of options;
- $\bar{T}$ is a state transition function such that $\bar{T}(\bar{s}, o, \bar{s}') = \sum_{s \in S, z(s)=\bar{s}} w_s \sum_{s' \in S, z(s')=\bar{s}'} P(s'|s, o)$ for any $\bar{s}, \bar{s}' \in \bar{S}$ and any option $o \in \bar{A}$;
- $\bar{R}$ is a reward function such that, for any $\bar{s} \in \bar{S}$ and any $o \in \bar{A}$, $\bar{R}(\bar{s}, o) = \sum_{s \in S, z(s)=\bar{s}} w_s R(s, o)$,

where w_s is the weight of state s, calculated based on the expected frequency of occurrence of state s in the abstract state $z(s)$ [11].

A *parameterised* AMDP uses parameters to specify which option to perform in each AMDP state [31]. An *abstract policy* fixes the values of all these parameters, and thus resolves the non-determinism of the AMDP, essentially transforming it into a Markov chain since there is a fixed, single option for each state.

3 Running Example

We will illustrate the application of ARL using an extension of the benchmark RL flag-collection mission from [24]. In the original mission, an agent learns to navigate a series of rooms and hallways in order to find and collect flags scattered throughout a building. In our extension, the building is augmented with security cameras that monitor certain doorways between areas. The detection of the agent by a camera results in the capture of the agent and the termination of its flag-collection mission. An illustration of this environment is shown in Fig. 1.

The agent is camouflaged, yet there is still a probability that it can be detected. The detection effectiveness of the cameras decreases towards the boundary of their field of vision, we represent this by sectioning the camera-monitored doorways into three areas: direct view by the camera, partial view, and hidden. These three areas are associated with decreasing probabilities of detection (Table 1).

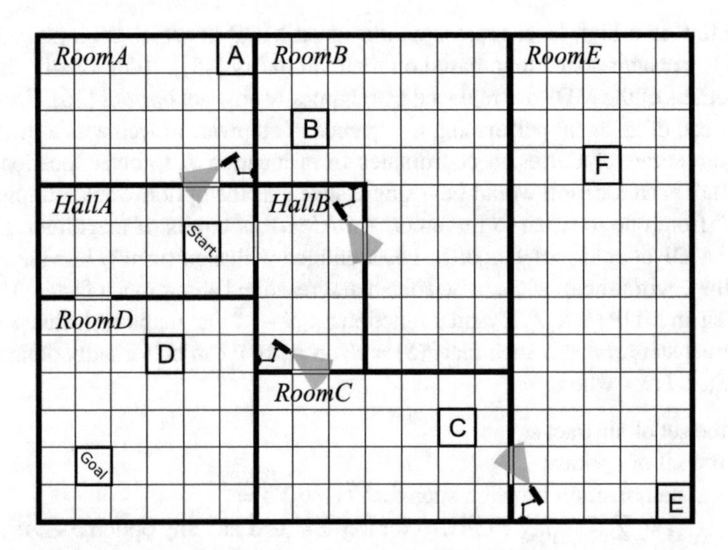

Fig. 1 Augmented flag-collection domain, showing the locations of the security cameras and their scope of vision, start and goal positions for the agent and flags to collect A-F

Table 1 Detection probabilities when transitioning between areas

Area transitions	View detection probabilities		
	Direct	Partial	Hidden
HallA ↔ RoomA	0.18	0.12	0.06
HallB ↔ RoomB	0.15	0.1	0.05
HallB ↔ RoomC	0.15	0.1	0.05
RoomC ↔ RoomE	0.21	0.14	0.07

Table 2 Constraints for the flag-collection mission

ID	Constraint	PCTL
C_1	The probability that the agent reaches the 'goal' area should be at least 0.75	$\mathrm{P}_{\geq 0.75}[\mathrm{F}\ goal]$
C_2	The agent should cumulate a reward above 2 before reaching the 'goal'	$\mathrm{R}_{>2}[\mathrm{F}\ goal]$

Suppose that in the real world, where the agent is actually a physical vehicle of some value, the owners of the vehicle wish for the safe return of it. However, they do not want it to behave "too safely" or it will not collect enough flags. Therefore, they specify the constraints from Table 2 for the agent. In this way, the right level of risk can be selected for each instance of the mission. Note that formulating the constraints C_1 and C_2 into a reward function and using standard RL to solve the problem is not possible since these objectives conflict and RL seeks to maximize a reward or minimize a cost, not maintain it within a specified range.

4 Method

As shown in Fig. 2, ARL takes as input a description of the problem to solve that comprises: (a) incomplete knowledge about the environment (i.e. problem); and (b) the set of constraints $C = \{C_1, C_2, \ldots, C_n\}$ that must be satisfied by the policy obtained by the RL agent. The incomplete knowledge must contain sufficient information for the *conservative* QV analysis of the environment properties associated with the $n > 0$ constraints. For instance, given the constraint C_1 from Table 2, it is sufficient to know a conservative lower bound for the detection probabilities of the cameras from the flag-collection mission in our running example. Note that the incomplete knowledge about the environment assumed by ARL is necessary: no constraints could be ensured during RL exploration in the absence of any information about the environment.

Under the assumptions detailed above, our ARL method yields a policy that satisfies the constraints C. To this end, ARL employs a process that integrates quantitative verification and reinforcement learning, and comprises the following stages:

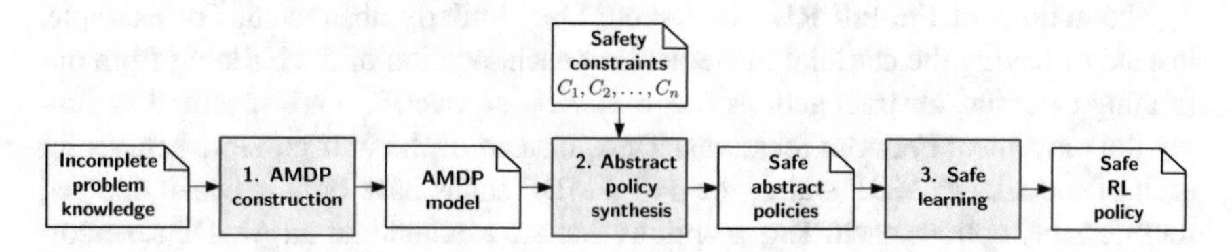

Fig. 2 Stages of the ARL method

1. **AMDP construction** – This ARL stage devises a parameterised AMDP model of the RL problem that supports the QV of the constraints $C_1, C_2, \ldots, C_n$.
2. **Abstract policy synthesis** – This stage generates AMDP policies (i.e. *abstract policies*) that satisfy the problem constraints. These "safe" policies are used to assemble an approximate Pareto-front of abstract policies. The optimisation objectives used to establish the Pareto dominance between different abstract policies are derived directly from the constraints (as described later in Sect. 4.2) or specified manually.
3. **Safe learning** – In this ARL stage, a suitable abstract policy from the Pareto-front is selected, and translated into state-action constraints for the exploration of the environment by the RL agent. Accordingly, the RL agent obtains an optimal policy that complies with the problem constraints.

4.1 Stage 1: AMDP Construction

In this ARL stage, all features that are relevant for the problem constraints must be extracted from the available incomplete knowledge about the RL environment. This could include locations, events, rewards, actions or progress levels. The objective is to abstract out the features that have no impact on the solution attributes that the constraints C refer to, whilst retaining as much detail as possible about the key features that these attributes depend on. This ensures that the constructed AMDP is sufficiently small to be analysed using quantitative verification, while also containing the necessary details to enable the analysis of all constraints.

In our running example, the key features are the locations and connections of rooms and halls, the detection probabilities of the cameras and the progress of the flags collected. Instead of having each (potentially unknown) Cartesian coordinate within a room or hall as a separate state, the room or hall as a whole is considered a single state in the AMDP. Also, we only consider the conservative detection probability per camera (which allows a conservative verification of constraint C_1 from Table 2), since the probabilities from Table 1 are unknown to the agent at this stage.

These abstractions reduce the size of the RL MDP, which is unknown to the agent and contains 14,976 states, to just 448 states for the associated AMDP. Note that the number of AMDP states is larger than the number of locations (i.e. rooms and halls) because different AMDP states are used for each possible combination of a location and a number of flags collected so far.

The actions of the full RL MDP should be similarly abstracted. For example, instead of having the cardinal movements at each location of the building from our running example, abstract actions (i.e. *options* – cf. Sect. 2.1) are specified as simply the movement between locations. Thus, instead of the four possible actions for each of the 14,976 MDP states, the 448 AMDP states have only between one and four possible options each. The m options that are available for an AMDP state correspond to the $m \geq 1$ passageways that link the location associated with that state with other locations, and can be encoded using a *state parameter* that takes one of the discrete values 1, 2, ..., m. The parameters for AMDP states with a single passageway (corresponding to rooms A, B and E from Fig. 1) can only take the value 1 and are therefore discarded. This leaves a set of 256 parameters that correspond to approximately 4×10^{99} possible abstract policies.

4.2 Stage 2: Abstract Policy Synthesis

In this ARL stage, a heuristic is used to find abstract policies that satisfy the constraints C for the AMDP constructed in Stage 1. The process is made easy by the use of the state parameters proposed in the previous section. Thus, each abstract policy can be obtained by assigning suitable values to these parameters. Fixing these parameter values in the AMDP resolves all nondeterminism, and the resulting model (which is a *Markov chain*) can be verified using QV, to establish if the abstract policy satisfies each constraint from C. If it does, the policy is deemed "safe", and is considered for inclusion in an approximate Pareto-optimal set of abstract policies. This set consists of abstract policies that *Pareto-dominate* each other according to a number of optimisation objectives such as probability of success or cumulated reward, where a policy π_A is said to Pareto-dominate another policy π_B iff π_A gives superior results to π_B for at least one objective, and for all other objectives π_A it is at least as good as π_B [32]. A Pareto-optimal set and the associated *Pareto-front* of objective values allow an acceptable trade-off between objectives to be determined a posteriori.

Algorithm 1 shows how this process of obtaining a Pareto-front of abstract policies is performed. The algorithm takes as its inputs the AMDP $\mathcal{M}$, the set of constraints C and the set of objectives to be optimised $\mathcal{O}$. Starting with the initially empty set of Pareto policies PS (line 2), a while loop begins which first generates a set of candidate abstract policies P using the function GETCANDIDATEPOLICIES (line 4). This function can be a search heuristic such as a genetic algorithm [33], hill climbing, simple random search or another search technique that may prove effective for the specific AMDP. Each policy $\pi \in P$ is then verified against each constraint $c \in C$ by the function PMC_1 which invokes a probabilistic model checker (e.g. PRISM) and returns *true* if the constraints are satisfied and *false* otherwise (line 6). If the policy satisfies all the constraints it is then compared to other safe policies (lines 8–15) to see if it Pareto-dominates any of them by function DOM (lines 24–26), where PMC_2 computes the value for the optimisation properties $o \in \mathcal{O}$. Those policies $\pi' \in PS$ which are pareto dominated by π are removed from the set, if π itself is not dominated by any other policies then it is included in PS. The outer while loop in lines 3–21 terminates once some criterion for $\neg\text{DONE}(PS)$ is fulfilled, for example if PS is sufficiently large or if a predefined number of iterations has been performed.

The optimisation objectives used to assess if either of two abstract policies Pareto-dominates the other can be specified manually or can be derived automatically from the constraints C. In the former case, additional PCTL formulae need to be formulated. In the latter case, the PCTL formula for each constraint C_i that specifies a lower bound for an attribute of the RL problem is interpreted as an attribute whose value

Algorithm 1 Abstract policy synthesis heuristic

```
1: function GENABSTRACTPOLICIES(M, C, O)
2:     PS ← {}
3:     while ¬DONE(PS) do
4:         P ← GETCANDIDATEPOLICIES(PS, M)
5:         for π ∈ P do
6:             if ⋀_{c∈C} PMC_1(M, π, c) then
7:                 dominated = false
8:                 for π' ∈ PS do
9:                     if DOM(π, π', M, O) then
10:                        PS ← PS \ {π'}
11:                    else if DOM(π', π, M, O) then
12:                        dominated = true
13:                        break
14:                    end if
15:                end for
16:                if ¬dominated then
17:                    PS ← PS ∪ {π}
18:                end if
19:            end if
20:        end for
21:    end while
22:    return PS
23: end function

24: function DOM(π_1, π_2, M, O)
25:     return
            ∀o∈O · PMC_2(M,π_1,o)≥PMC_2(M,π_2,o)∧
            ∃o∈O · PMC_2(M,π_1,o)>PMC_2(M,π_2,o)
26: end function
```

should be maximised. Conversely, attributes for which upper bounds are specified in the problem constraints are considered attributes whose value should be minimised.

In our running example, the two constraints for the flag-collection problem specify lower bounds both for the probability that the agent reaches the 'goal' area and for the reward cumulated by the RL agent. Therefore, using automated selection of optimisation objectives for the running example yields an approximate Pareto-front corresponding to these two attributes being maximized.

4.3 Stage 3: Safe Learning

The last ARL stage exploits the approximate Pareto-optimal set of abstract policies synthesised in Stage 2. A policy is selected from this set by taking into account the trade-offs that different policies achieve for the optimisation objectives used to assemble the set. The selected abstract policy is then used to ensure that the RL agent achieves the required constraints by removing the low-level MDP actions that do not correspond to *options* from the abstract policy. For instance, assume that the selected abstract policy for our running example requires the agent to never enter RoomA. In this case, should the agent be at Cartesian coordinates (5, 9) (i.e. the position immediately to the North of the Start position), the action to move North and thus to enter RoomA is removed from the agent's action set, for this specific state. By disallowing the actions that are not associated with options permitted by the abstract policy, the RL agent's learning and learnt low-level behaviours are guaranteed to satisfy the problem constraints, as illustrated in the next section for a case study.

Abstract policies intentionally reduce agent autonomy to prevent unsafe actions, but do not preclude it completely. For example, in the running example the agent must learn the flag locations within the rooms as well as the doorway areas safest to cross, information which is not contained within the abstract policies. Whilst abstract policy constraints may yield suboptimal RL policies with respect to the RL model in its entirety, this key feature guarantees safety.

5 Evaluation

5.1 Experimental Setup

We evaluated the effectiveness and generality of our ARL approach by applying it to two case studies. The first case study is based on the navigation task described in Sect. 3, where the learning agent must navigate a guarded environment comprising hallways and rooms in order to collect flags distributed throughout. Our second case study is adapted from [25], where an assisted-living system has been developed to aid dementia sufferers with daily living tasks.

For each case study we conducted a set of four experiments. The first of these experiments did not involve the use of our ARL approach, and was a standard RL implementation of the case study. This experiment serves as a baseline which we contrast with the ARL experiments in order to determine the effects of our method.

All experiment parameters were chosen empirically in line with standard RL practice. For all experiments we use a discount factor $\gamma = 0.99$ and a learning rate $\alpha = 0.1$ which decays to 0 over the learning run. Parameters specific to individual experiments are mentioned throughout this section. As is convention when evaluating stochastic processes, we repeated each experiment multiple times (i.e. five times) and we evaluated the final policy for each experiment many times (i.e. 10,000 times) in order to ensure that the results are suitably significant [34].

Learning evaluation is done after each learning episode during a run, whilst we only perform five learning runs per experiment, error bars for the standard error of the mean show the statistical significance of the learning (Figs. 4, 5, 8 and 9). Policy evaluations were done once learning had finished (Tables 4 and 8).

We implemented the ARL experiments using the York Reinforcement Learning Library (YORLL, www.cs.york.ac.uk/rl/software.php), which supports a wide range of environments and learning algorithms. For the QV component of ARL, we used the PRISM model checker [27], which supports the verification of reward-extended PCTL properties for MDPs and has been successfully used to analyse similar models of systems ranging from cloud infrastructure [35] and service-based systems [36] to unmanned vehicles [37, 38].

5.2 *Guarded Flag-Collection*

This case study is based on the scenario described in Sect. 3 and referred to throughout Sect. 4. In the interest of brevity, the details presented in these two previous sections will not be repeated here.

In our RL implementation, the agent receives a reward of 1 for each flag it collects and an additional reward of 1 for reaching the 'goal' area of the building. If the agent is captured the agent receives a reward of -1, regardless of any flags already collected.

We used the AMDP constructed during the first ARL stage as described in Sect. 4.1. In the second ARL stage, we generated 10,000 abstract policies with parameter values (i.e. state to action mappings) drawn randomly from a uniform distribution. Out of these abstract policies, QV using the probabilistic model checker PRISM identified 14 policies that satisfied the two constraints from Table 2. Figure 3 shows the QV results obtained for these 14 abstract policies, i.e. their associated probability of reaching the 'goal' area and expected number of flags collected. The approximate Pareto-front depicted in this figure was obtained using the two optimization

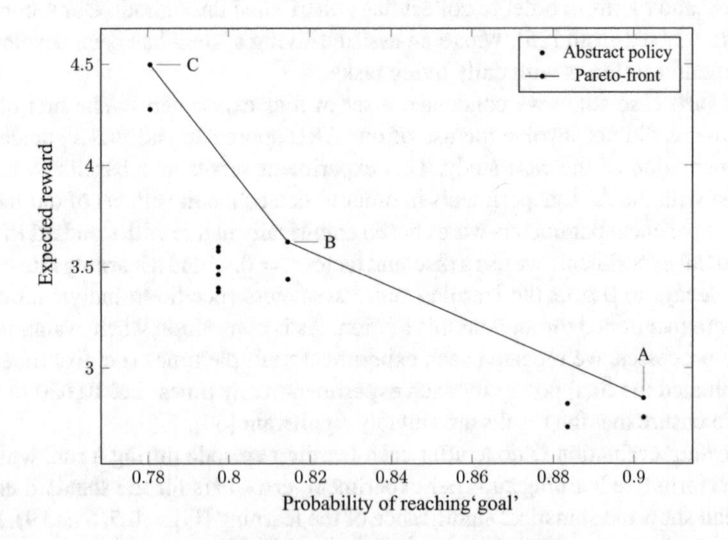

Fig. 3 Pareto-front of abstract policies that satisfy the constraints from Table 2. Those policies that were selected for ARL are labelled A, B and C

Table 3 Selected abstract policies to use for ARL in the guarded flag-collection

Abstract policy	Probability of reaching 'goal'	Expected reward
A	0.9	2.85
B	0.81	3.62
C	0.78	4.5

Table 4 Results for baseline and ARL experiments for guarded flag-collection

Abstract Policy	Probability of reaching 'goal'	Standard error	Expected reward	Standard error
None	0.72	0.0073	4.01	0.031
A	0.9	0.0012	2.85	0.0029
B	0.81	0.0019	3.62	0.0037
C	0.78	0.0012	4.5	0.0041

objectives described in Sect. 4.2, i.e. maximizing the expected number of flags collected and the probability of reaching the 'goal' area of the building.

From this Pareto-front we selected three abstract policies (labelled as A, B and C in Fig. 3) to use in different experiments during the safe learning ARL stage, as explained in Sect. 4.3. The properties of these three abstract policies are shown in Table 3.

The baseline experiment (which did not use ARL) used an exploration probability $\epsilon = 0.8$ and performed 2×10^7 learning episodes, each with 10,000 steps. This did not, however, reach a global optimum. Even after extensive learning, in excess of 10^9 learning episodes, conventional RL did not attain a superior solution. In contrast to our experiments where ARL was used, cf. abstract policy C, Table 4, a superior policy was learned much faster, further demonstrating the advantages of our approach. Figure 4 shows the learning progress for this experiment.

Next, we ran three further sets of RL experiments, one for each of the abstract policies from Table 3. It was not necessary to have so many learning episodes as for the baseline experiment, since the abstract policy had the effect of guiding the agent with regard to the locations to enter next, and therefore only 10^5 episodes were necessary for the learning to converge. Similarly, a lower exploration probability of $\epsilon = 0.6$ proved best since the abstract policy reduced the state space the agent needed to explore. Figure 5 show the RL learning progress for each of the abstract policies used for ARL.

As can be seen from the results summarized in Table 4, the experiments where an abstract policy was applied resulted in an RL policy that: (a) satisfied the problem constraints from Table 2; and (b) matched the probability of reaching the 'goal' area and the expected reward of the abstract policy (cf. Table 3). The baseline experiment gave results that do not satisfy our constraints, which was expected given that only 14 of the 10,000 abstract policies synthesised by ARL satisfied these constraints.

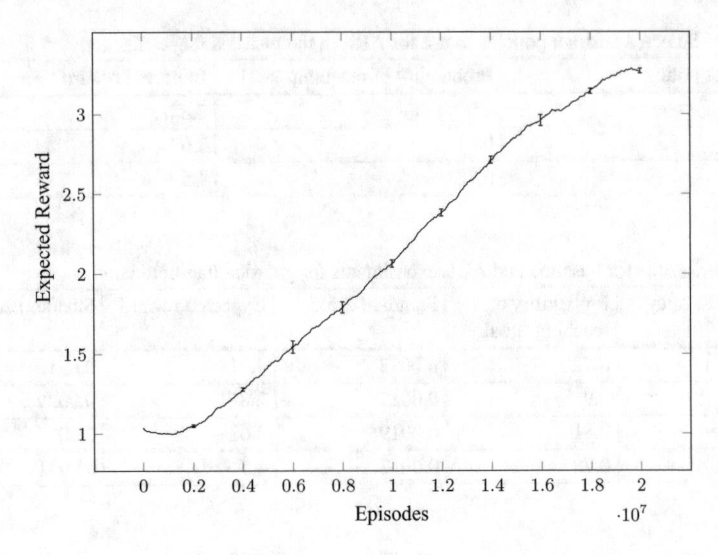

Fig. 4 Learning for guarded flag-collection with no ARL applied

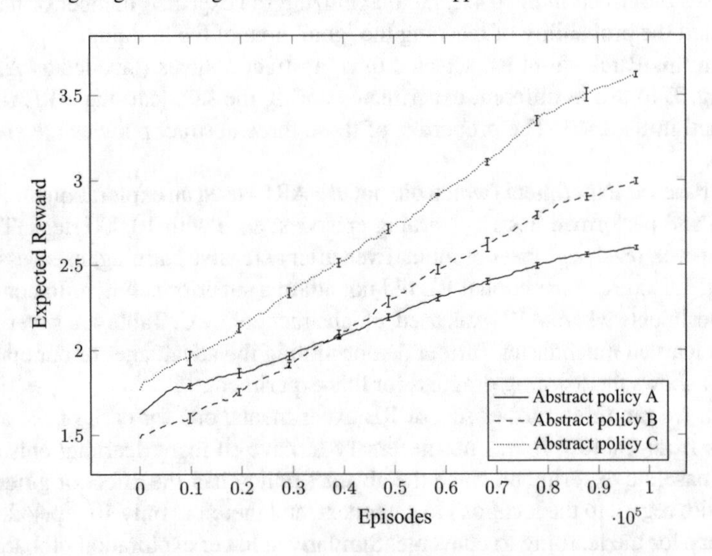

Fig. 5 Learning for guarded flag-collection with ARL applied using the selected abstract policies A, B and C

5.3 Assisted-Living System for Dementia Sufferer

Dementia is an illness where sufferers experience, amongst other symptoms, the decreasing ability to perform daily tasks. For example, the task of washing ones hands can present as a significant challenge as the sufferer can struggle to recollect

Table 5 The individual subtasks for washing hands and their associated the boolean atomic propositions indicating if the task has been performed

Subtask	Atomic proposition
Turn tap on	on
Apply soap	soaped
Wet hands under tap	wet
Rinse washed hands	rinsed
Dry hands	dried

the correct order of events they must perform to complete the task, or they may forget what stage of the process they have reached so far and as a consequence may regress by repeating things they have already done. It is therefore necessary for someone, e.g. a nurse, to care for the sufferer by assisting them with such tasks.

The constant need for a caregiver to be present, however, can be a considerable burden to them, furthermore, can be costly for the health service. Therefore, a system has been designed in [25] to substitute for a caregiver to help perform these daily tasks. The "assisted-living system" uses an MDP to model the task of hand-washing and in certain states the system can give a voice prompt to the sufferer with the aim to guide them on what they must do next, if they are struggling to progress. These voice prompts become increasingly explicit in their instructions if the sufferer repeatedly fails to progress. Should the sufferer still fail to progress after three consecutive prompts the system can resort to summoning the caregiver to intervene.

The task of hand-washing can be broken down into several subtasks. Table 5 lists each of the subtasks involved when washing ones hands. Throughout this section we will use boolean atomic propositions to represent whether or not the subtask has been achieved or not. Fig. 6 illustrates the workflow when washing hands, including the possible progressions and regressions that can be made by a dementia sufferer.

We have adapted this system for use as a second case study. In our RL implementation of the system we have expanded it by introducing different genders

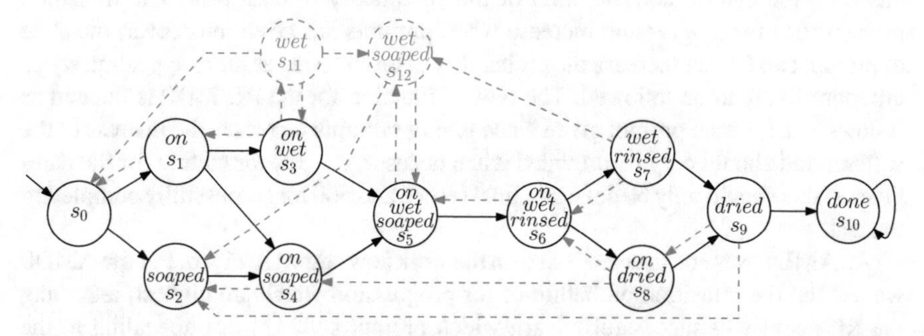

Fig. 6 The subtasks involved when washing hands. Progress is shown as black, continuous-lines and nodes and the possible regressions as red, dashed-lines and nodes. Nodes where atomic propositions are present indicates that they are *true*, where they do not appear they are *false*

Table 6 Constraints and optimisation objectives for the assisted-living system where m is the number of mistakes made at any given time, $MAX\_MISTAKES$ is the maximum allowed number of mistakes before summoning the caregiver, distress is the reward structure for number of prompts given and "done" represents completion of the task

ID	Constraint (C)/Optimisation objective (O)	PCTL
C_1	The probability that the caregiver provides assistance should be at least 0.05	$P_{\geq 0.05}[F\, m = MAX\_MISTAKES]$
C_2	The probability that the caregiver provides assistance should be at most 0.2	$P_{\leq 0.2}[F\, m = MAX\_MISTAKES]$
O_1	The level of dementia sufferer distress due to multiple voice prompts should be minimised	minimise $R^{\text{distress}}_{=?}[F\, done \vee m = MAX\_MISTAKES]$
O_2	The probability of calling the caregiver should be minimised (subject to C_1 and C_2 being satisfied)	minimise $P_{=?}[F\, m = MAX\_MISTAKES]$

(male/female) and volumes (soft/moderate/loud) for the voice prompt, where the RL agent learns which gender/volume is most conducive to the agent progressing.

The RL agent must also learn when to give prompts so to (a) minimize the probability that the caregiver is called for, whilst (b) minimizing the number of prompts given to the sufferer since persistent prompts can become distressing to the patient. In addition to these optimisation objectives, we also desire that the caregiver must be present sometimes so that the sufferer does not feel neglected, yet we do not wish them to be present too frequently which would defeat the purpose of the system. Therefore, supposing a person washes there hands approximately five times a day, and we wish the caregiver to be present at least once every one-to-four days, the probability of them intervening should be between 0.05 and 0.2. These optimisation objectives and constraints are specified as PCTL formulae in Table 6.

The transition probabilities between each stage of the task were approximated based on the careful consideration of the complexity of each task. The transition probabilities for progressing increase when prompts are given, moreover, the style of prompt can further increase the probability of progressing as more appealing styles are more likely to be followed. The reward function for the RL MDP is defined as follows: -1 for each prompt given since giving prompts increases the distress to the sufferer and should only be provided when necessary, -300 for calling for the caregiver as this should only be done as a final resort, and 500 for successfully completing the task.

An AMDP was constructed based on the workflow shown in Fig. 6. For the AMDP we use the the transition probabilities for progression which are highest, assuming the RL agent will successfully learn which prompt style is most appealing to the sufferer. The reward function for the AMDP only requires -1 for when prompts are given; rewards for completing the task and for calling the caregiver are only required

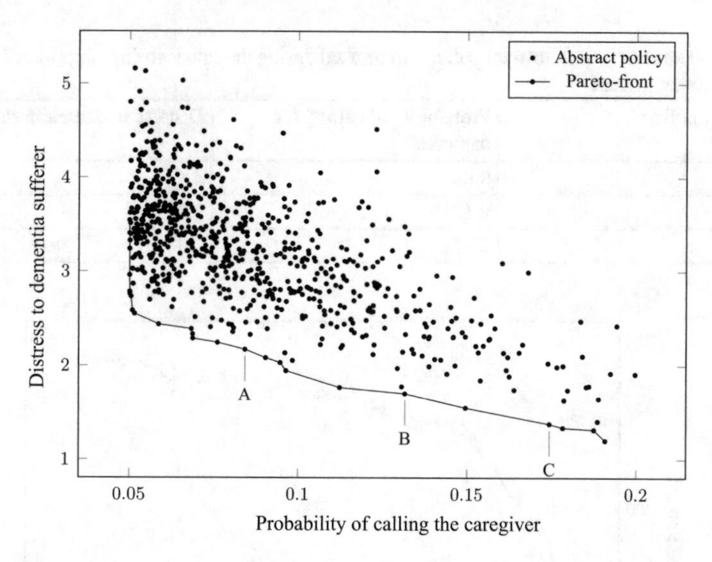

Fig. 7 Safe abstract policies for the assisted-living system including the Pareto-front. The three policies chosen for ARL are labelled as A, B and C

by the RL agent so it learns to reach the "done" absorbing state s_{10} as well as to not immediately call the caregiver, respectively.

Abstract policies for the AMDP contain 12 parameters, once for each stage of the workflow excluding s_{10} where the task has been completed and no prompts are required. Each parameter defines the maximum number of mistakes for each stage that are necessary for a prompt to eventually be given, where a parameter value of 0 will mean that a prompt is always given, and a value equal to the maximum mistakes threshold before calling the caregiver will mean never giving a prompt. An additional parameter to find is the threshold for the maximum number of mistakes allowed before calling for the caregiver.

As with the previous case study, we generated 10,000 candidate abstract policies, of which 786 satisfied the constraints C_1 and C_2 from Table 6. The Pareto-front of these safe abstract policies was generated subject to optimisation objectives O_1 and O_2 from Table 6. Figure 7 shows this Pareto-front with three policies chosen for ARL labelled A, B and C. The probability of calling the caregiver and level of distress to the dementia sufferer for these three policies are listed in Table 7.

For all experiments for this case study we used an $\epsilon = 0.5$ and episodes had a maximum of 1,000 steps. For the baseline experiment 10^6 episodes were necessary, more were required for the safe RL experiments since the nature of the abstract policies to sometimes prevent the agent from giving prompts meant the agent was unable to try them as frequently, increasing the time needed to learn which prompt style was best. The learning progress for the baseline experiment is shown in Fig. 8, and for the experiments using safe abstract policies is shown in Fig. 9.

Table 7 The chosen safe abstract policies to be used during the safe learning stage of ARL for the assisted-living system

Abstract policy	Probability of calling the caregiver	Distress to dementia sufferer
A	0.08	2.17
B	0.13	1.70
C	0.17	1.38

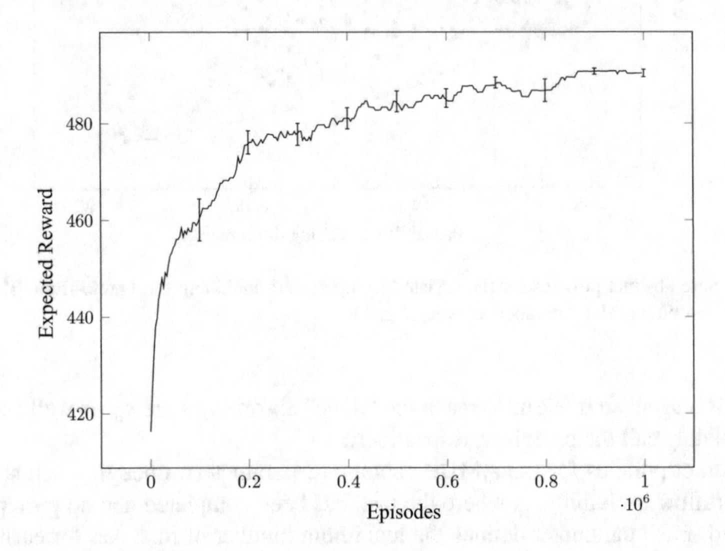

Fig. 8 Learning for assisted-living system with no ARL applied

The results from these experiments are summarised in Table 8. As can be seen, the safe RL experiments (using abstract policies A, B and C) satisfy the safety requirements specified in Table 6, furthermore, the baseline RL experiment without safe learning did not. Comparing with the policy attributes from Table 7, the results closely match the probability of calling the caregiver and the distress to the patient. The probability of calling the caregiver for safe learning using policy C is slightly higher than that policy's probability in Table 7, this can be attributed to the fact that the learned policy was not completely optimal and with extra learning episodes should converge to the same probability.

6 Related Work

The ARL technique introduced in our paper belongs to a class of RL techniques called safe reinforcement learning [7]. The previous research on safe RL has proposed techniques that can enforce bounds on the either the reward obtained by the

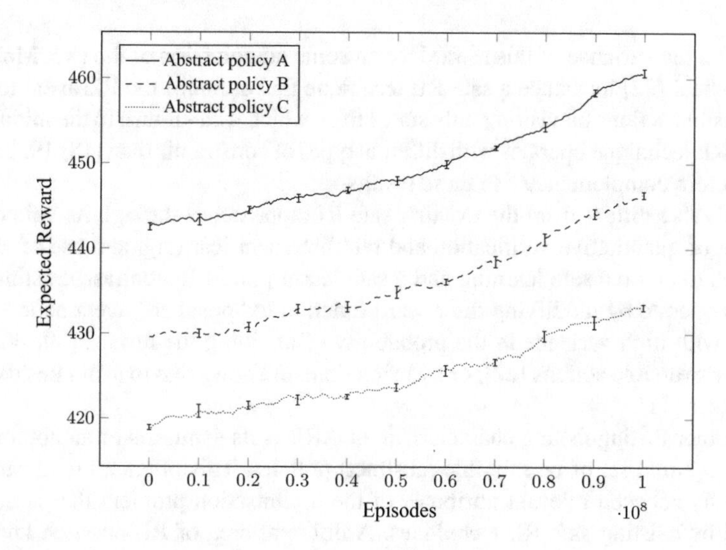

Fig. 9 Learning for assisted-living system with ARL applied using the selected abstract policies A, B and C

Table 8 Results from baseline and ARL experiments for the assisted-living system

Abstract policy	Probability of calling the caregiver	Standard error	Distress to patient	Standard error
None	4.02×10^{-4}	4.28×10^{-4}	8.31	4.03×10^{-3}
A	0.08	4.95×10^{-4}	2.17	3.25×10^{-3}
B	0.13	5.17×10^{-4}	1.70	2.22×10^{-3}
C	0.18	4.27×10^{-4}	1.38	1.84×10^{-3}

RL agent or on simple measures that are related to its optimisation. Alternative approaches modify how the agent explores the environment to avoid areas of risk.

The technique proposed by Geibel [21] supports an inequality constraint on the reward cumulated by the RL agent or a maximum permitted probability for such a constraint to be violated. Safe RL techniques that support similar constraints by generalizing chance-constrained planning to infinite-horizon MDPs are presented by Mannor and Delage [20] and Ponda et al. [23]. The constraints supported by [20, 21, 23] are a subset of the types of constraints supported by ARL, which can handle the wide range of constraints that can be specified in PCTL.

Abe et al. [18] describe a safe RL technique in which high-level business and legal "constraint rules" are enforced during each value iteration of the RL process, and apply it to a tax collection optimization problem. Building on insights from financial decision making and robust process control, Castro et al. [19] introduce a safe RL technique that enforces constraints on the cumulative reward obtained by the RL agent, on the variance of this reward, or on some combination of the two. Moldovan and Abbeel [22] introduce a safe RL technique that enforces the RL agent to avoid irreversible actions by visiting only states from which it can return to the initial state. Our ARL technique operates with different types of constraints than [18, 19, 22], and is therefore complementary to these results.

ARL also differs from the existing safe RL approaches through its unique integration of quantitative verification and reinforcement learning, and use of abstract policies to enforce safe learning and a safe learnt policy. In contrast, existing techniques operate by modifying the reward function to "penalize" agent actions associated with high variance in the probability of attaining the reward [39, 40] or to avoid irreversible actions [22], or by using domain knowledge to avoid unsafe states altogether [41].

Another distinguishing characteristic of ARL is its synthesis of an approximate Pareto-optimal set of permissible (abstract) policies. This offers a broad choice of trade-offs between relevant attributes of the optimisation problem that is not supported by existing safe RL techniques. A different area of RL research known as multi-objective RL (MORL) [32, 42] has also considered the problem of learning a policy that satisfies multiple objectives that may conflict with each other. However, neither single-policy MORL algorithms (which learn an optimal policy for each objective and then combine them to form a single policy [43, 44]) nor multi-policy algorithms (which learn an approximate Pareto-front for each objective [45] or a joint Pareto-front [46]) support the rich expressiveness provided by ARL through its use of reward-augmented PCTL constraints.

7 Conclusion

We proposed the use of an abstract MDP formally analysed using quantitative verification as a means to restrict the action set of an RL agent to the actions that were proven to satisfy a set of required constraints, adding to the growing research on safe RL. Through two case studies based on the benchmark RL flag-collection domain and an assissted-living planning system, we demonstrated that the hybrid ARL technique can be applied successfully. ARL requires that partial knowledge of the problem is provided a priori, and makes the typical assumption that RL will converge towards an optimal policy.

Unlike standard RL, our technique supports a wide range of safety, performance and reliability constraints that cannot be expressed using a single reward function and are not supported by existing safe RL techniques. Furthermore, the use of an AMDP allows the application of ARL with only limited knowledge about the environment, and ensures that ARL scales to much larger and complex models than would otherwise be feasible. Additionally, the construction of the AMDP and the use of PCTL formulae, an expressive and convenient representation formalism for required properties, enables on the fly experimentation of constraints and properties without requiring modification of the underlying model.

Our future work on the ARL technique will include exploiting some of the more sophisticated constraints that can be specified in PCTL. For example, unbounded until PCTL formulae can be used to constrain the order in which the agent visits different rooms in the guarded flag-collection case study, e.g. $P_{\geq 0.9}[\neg RoomA \; \mathrm{U} \, \boldsymbol{RoomB}]$ requires that, with a probability of at least 0.9, the agent should not visit RoomA before RoomB. Furthermore, bounded until PCTL formulae can additionally place constraints on the number of time steps taken to achieve a certain outcome.

Additionally, we plan to research a means of updating the AMDP should it not accurately reflect the RL MDP. As we mention in [47, 48], in the event that the RL agent encounters dynamics in the RL MDP that do not correlate with the AMDP, a means of feeding back this information to update the AMDP can be developed based on [49–51]. After updating the AMDP the constraints will need to be reverified and, if necessary, a new abstract policy will be generated.

Finally, we also plan to explore the possibility to employ multi-objective RL [32, 42] in the abstract policy synthesis of ARL, and to develop a variant of the technique where the actions restricted through QV and those learnt by the RL agent belong to disjoint sets. This ARL variant will support the common scenario in which a set of system attributes like cost and energy usage need to be optimised once strict constraints are guaranteed to be satisfied for some other system attributes such as availability and response time.

Acknowledgements This paper presents research sponsored by the UK MOD. The information contained in it should not be interpreted as representing the views of the UK MOD, nor should it be assumed it reflects any current or future UK MOD policy.

References

1. Wiering, M., Otterlo, M.: Reinforcement learning and markov decision processes. In: Wiering, M., Otterlo, M. (eds.) Reinforcement Learning: State-of-the-art, pp. 3–42. Springer, Berlin, Heidelberg (2012)
2. Riedmiller, M., Gabel, T., Hafner, R.: Reinforcement learning for robot soccer. Auton. Robots **27**(1), 55–73 (2009)
3. Kwok, C., Fox, D.: reinforcement learning for sensing strategies. In: IEEE/RSJ International Conference on Intelligent Robots and Systems, pp. 3158–3163. IEEE (2004)
4. Baxter, J., Tridgell, A., Weaver, L.: KnightCap: A chess program that learns by combining TD(λ) with minimax search. In: 15th International Conference on Machine Learning, pp. 28–36. Morgan Kaufmann (1998)
5. Bagnell, J.A., Schneider, J.G.: autonomous helicopter control using reinforcement learning policy search methods. In: IEEE International Conference on Robotics and Automation. IEEE (2001)
6. Calinescu, R., Ghezzi, C., Kwiatkowska, M., Mirandola, R.: Self-adaptive software needs quantitative verification at runtime. Commun. ACM **55**(9), 69–77 (2012)
7. García, J., Fernández, F.: A comprehensive survey on safe reinforcement learning. J. Mach. Learn. Res. **16**(1), 1437–1480 (2015)

8. Mason, G., Calinescu, R., Kudenko, D., Banks, A.: Combining reinforcement learning and quantitative verification for agent policy assurance. In: 6th International Workshop on Combinations of Intelligent Methods and Applications, pp. 45–52 (2016)
9. Kwiatkowska, M.: Quantitative verification: models, techniques and tools. In: 6th Joint Meeting of the European Software Engineering Conference and the ACM SIGSOFT Symposium on the Foundations of Software Engineering, pp. 449–458. ACM Press (2007)
10. Hansson, H., Jonsson, B.: A logic for reasoning about time and reliability. Formal Aspects Comput. **6**(5), 512–535 (1994)
11. Marthi, B.: Automatic shaping and decomposition of reward functions. In: 24th International Conference on Machine learning, pp. 601–608. ACM (2007)
12. Lahijanian, M., Andersson, S.B., Belta, C.: Temporal logic motion planning and control with probabilistic satisfaction guarantees. IEEE Trans. Robot. **28**(2), 396–409 (2012)
13. Cizelj, I., Ding, X.C. and Lahijanian, M., Pinto, A., Belta, C.: Probabilistically safe vehicle control in a hostile environment. In: 18th World Congress of the The International Federation of Automatic Control, pp. 11803–11808. (2011)
14. Feng, L., Wiltsche, C., Humphrey, L., Topcu, U.: Synthesis of human-in-the-loop control protocols for autonomous systems. IEEE Trans. Autom. Sci. Eng. **13**(2), 450–462 (2016)
15. Li, L., Walsh, T.J., Littman, M.L.: Towards a unified theory of state abstraction for MDPs. In: 9th International Symposium on Artificial Intelligence and Mathematics, pp. 531–539 (2006)
16. Sutton, R.S., Precup, D., Singh, S.: Between MDPs and Semi-MDPs: a framework for temporal abstraction in reinforcement learning. Artif. Intell. **112**(1–2), 181–211 (1999)
17. Andova, S., Hermanns, H., Katoen, J.-P.: Discrete-time rewards model-checked. In: Larsen, K.G., Niebert, P. (eds) Formal Modeling and Analysis of Timed Systems. LNCS, vol. 2791, pp. 88–104 (2004)
18. Abe, N., Melville, P., Pendus, C., Reddy, C.K., Jensen, D.L., Thomas, V.P., Bennett, J.J., Anderson, G.F., Cooley, B.R., Kowalczyk, M., Domick, M., Gardinier, T.: Optimizing debt collections using constrained reinforcement learning. In: 16th ACM SIGKDD International Conference on Knowledge Discovery and Data Mining, pp. 75–84. ACM (2010)
19. Castro, D.D., Tamar, A., Mannor, S.: Policy gradients with variance related risk criteria. In: 29th International Conference on Machine Learning, pp. 935–942. ACM (2012)
20. Delage, E., Mannor, S.: Percentile optimization for markov decision processes with parameter uncertainty. Oper. Res. **58**(1), 203–213 (2010)
21. Geibel, P.: Reinforcement learning for MDPs with constraints. In: Fürnkranz, J., Scheffer, T., Spiliopoulou, M. (eds.) 17th European Conference on Machine Learning. LNAI vol. 4212, pp. 646–653. Springer, Heidelberg (2006)
22. Moldovan, T.M., Abbeel, P.: Safe exploration in markov decision processes. In: 29th International Conference on Machine Learning, pp. 1711–1718. ACM (2012)
23. Ponda, S.S., Johnson, L.B., How, J.P.: Risk allocation strategies for distributed chance-constrained task allocation. In: American Control Conference, pp. 3230–3236. IEEE (2013)
24. Dearden, R., Friedman, N., Russell, S.: Bayesian Q-learning. In: 15th National Conference on Artificial Intelligence, pp. 761–768. AAAI Press (1998)
25. Boger, J., Hoey, J., Poupart, P., Boutilier, C., Fernie, G., Mihailidis, A.: A planning system based on markov decision processes to guide people with dementia through activities of daily living. IEEE Trans. Inf. Technol. Biomed. **10**(2), 323–33 (2006)
26. Kwiatkowska, M.: advances in quantitative verification for ubiquitous computing. In: Liu, Z., Woodcock, J., Zhu, H. (eds.) 10th International Colloquium on Theoretical Aspects of Computing. LNCS vol. 8049, pp. 42–58. Springer (2013)
27. Kwiatkowska, M., Norman, G., Parker, D.: PRISM 4.0: verification of probabilistic real-time systems. In: Gopalakrishnan, G., Qadeer, S. (eds.) 23rd International Conference on Computer Aided Verification. LNCS vol. 6806, pp. 585–591. Springer (2011)
28. Katoen, J.-P., Zapreev, I.S., Hahn, E.M., Hermanns, H., Jansen, D.N.: The Ins and Outs of the probabilistic model checker MRMC. Perform. Eval. **68**(2), 90–104 (2011)

29. Kwiatkowska, M., Norman, G., Parker, D.: Stochastic model checking. In: Bernardo, M., Hillston, J. (eds.) 7th International Conference on Formal Methods for Performance Evaluation. LNCS vol. 4486, pp. 220–270. Springer (2007)
30. Watkins, C.J.C.H., Dayan, P.: Q-Learning. Mach. Learn. **8**(3), 279–292 (1992)
31. Xia, L., Jia, Q.-S.: Policy Iteration for parameterized markov decision processes and its application. In: 9th Asian Control Conference, pp. 1–6. IEEE (2013)
32. Liu, C., Xu, X., Hu, D.: Multiobjective reinforcement learning: a comprehensive overview. IEEE Trans. Syst. Man Cybern. Syst. **45**(3), 385–398 (2015)
33. Gerasimou, S., Tamburrelli, G., Calinescu, R.: Search-based synthesis of probabilistic models for quality-of-service software engineering. In: 30th IEEE/ACM International Conference on Automated Software Engineering, pp. 319–330. IEEE (2015)
34. Arcuri, A., Briand, L.: A practical guide for using statistical tests to assess randomized algorithms in software engineering. In: 33rd International Conference on Software Engineering, pp. 1–10. ACM (2011)
35. Calinescu, R., Kikuchi, S., Johnson, K.: Compositional reverification of probabilistic safety properties for large-scale complex IT systems. In: Large-Scale Complex IT Systems. Development, Operation and Management, pp. 303–329. Springer (2012)
36. Calinescu, R., Johnson, K., Rafiq, Y.: Developing self-verifying service-based systems. In: 28th IEEE/ACM International Conference on Automated Software Engineering (ASE), pp. 734–737 (2013)
37. Calinescu, R., Gerasimou, S., Banks, A.: Self-adaptive software with decentralised control loops. In: 17th International Conference on Fundamental Approaches to Software Engineering, pp. 235–251. Springer (2015)
38. Gerasimou, S., Calinescu, R., Banks, A.: Efficient runtime quantitative verification using caching, lookahead, and nearly-optimal reconfiguration. In: 9th International Symposium on Software Engineering for Adaptive and Self-Managing Systems (SEAMS), pp. 115–124 (2014)
39. Heger, M.: Consideration of risk in reinforcement learning. In: 11th International Conference on Machine Learning, pp. 105–111. (1994)
40. Mihatsch, O., Neuneier, R.: Risk-sensitive reinforcement learning. Mach. Learn. **49**(2), 267–290 (2002)
41. Driessens, K., Džeroski, S.: Integrating Guidance into Relational Reinforcement Learning. Mach. Learn. **57**(3), 271–304 (2004)
42. Vamplew, P., Dazeley, R., Berry, A., Issabekov, R., Dekker, E.: Empirical evaluation methods for multiobjective reinforcement learning algorithms. Mach. Learn. **84**(1), 51–80 (2011)
43. Gábor, Z., Kalmár, Z., Szepesvári, C.: Multi-criteria reinforcement learning. In: 15th International Conference on Machine Learning, pp. 197–205. Morgan Kaufmann (1998)
44. Mannor, S., Shimkin, N.: A geometric approach to multi-criterion reinforcement learning. J. Mach. Learn. Res. **5**, 325–360 (2004)
45. Barrett, L., Narayanan, S.: learning all optimal policies with multiple criteria. In: 25th International Conference on Machine learning, pp. 41–47. Omni Press (2008)
46. Moffaert, K.V., Nowé, A.: Multi-objective reinforcement learning using sets of pareto dominating policies. J. Mach. Learn. Res. **15**(1), 3663–3692 (2014)
47. Mason, G., Calinescu, R., Kudenko, D., Banks, A.: assured reinforcement learning for safety-critical applications. In: Doctoral Consortium on Agents and Artificial Intelligence (DCAART '17), pp. 9–16 (2017)
48. Mason, G., Calinescu, R., Kudenko, D., Banks, A.: Assured reinforcement learning with formally verified abstract policies. In: 9th International Conference on Agents and Artificial Intelligence, pp. 105–117 (2017)
49. Calinescu, R., Johnson, K., Rafiq, Y.: Using observation ageing to improve Markovian model learning in QoS engineering. In: 2nd International Conference on Performance Engineering, pp. 505–510 (2011)

50. Calinescu, R., Rafiq, Y., Johnson, K., Bakir, M.E.: Adaptive model learning for continual verification of non-functional properties. In: 5th International Conference on Performance Engineering, pp. 87–98 (2014)
51. Efthymiadis, K., Kudenko, D.: Knowledge revision for reinforcement learning with abstract MDPs. In: Bordini, R.H., Elkind, E., Weiss, G., Yolum, P. (eds.) 14th International Conference on Autonomous Agents and Multiagent Systems, pp. 763–770. (2015)

Distillation of Deep Learning Ensembles as a Regularisation Method

Alan Mosca and George D. Magoulas

Abstract Ensemble methods are among the most commonly utilised algorithms that construct a group of models and combine their predictions to provide improved generalisation. They do so by aggregating multiple *diverse* versions of models learned using machine learning algorithms, and it is this diversity that enables the ensemble to perform better than any of its members taken individually. This approach can be extended to produce ensembles of deep learning methods that combine various good performing models, which are between them very diverse because they have reached different local minima and make different prediction errors. It has been shown that a large, cumbersome deep neural network can be approximated by a smaller network through a process of distillation, and that it is possible to approximate an ensemble of other learning algorithms by using a single neural network, with the help of additional artificially generated pseudo-data. We extend this work to show that an ensemble of deep neural networks can indeed be approximated by a single deep neural network with size and capacity equal to the single ensemble member, and we develop a recipe that shows how this can be achieved without using any artificial training data or any other special provisions, such as using the *soft output targets* during the distillation process. We also show that, under particular circumstances, the distillation process can be used as a form of regularisation, through its implicit reduction in learning capacity. We corroborate our findings with an experimental analysis on some common benchmark datasets in computer vision and deep learning.

Keywords Ensembles · Deep learning · Distillation · Convolutional neural networks

---

A. Mosca (✉) · G.D. Magoulas
Department of Computer Science and Information Systems Birkbeck,
University of London, London, UK
e-mail: a.mosca@dcs.bbk.ac.uk

G.D. Magoulas
e-mail: gmagoulas@dcs.bbk.ac.uk

I. Hatzilygeroudis and V. Palade (eds.), *Advances in Hybridization of Intelligent Methods*, Smart Innovation, Systems and Technologies 85,
https://doi.org/10.1007/978-3-319-66790-4_6

1 Introduction

A typical trait among almost all of the many available machine learning ensemble methods, of which a survey can be found in [6], is that at the time of testing (sometimes also called "inference time"), the input is run through every single member of the ensemble, in order to obtain the combined result. This makes the operation of the ensemble at test–time expensive when compared to each base classifier taken individually, with both time and memory requirements increasing linearly w.r.t. the number of members in the ensemble. Because of this cost relationship to the ensemble size, which is also mirrored in the training phase, ensembles have been usually only used with small base learner algorithms, or as a final addition to an independently fine–tuned, more complex, base learner. This means that deep learning ensembles have been largely overlooked as a field of research because of the high costs involved.

A process called *distillation* has been developed in [9], which shows how a reduced–complexity deep neural network is able to approximate the behaviour of a more *cumbersome* network. This is achieved by using the *soft output values* of the cumbersome network as a training target for the approximate network. In [4], the authors show that a simple feed-forward, fully-connected Artificial Neural Network is able to learn the distilled knowledge of an ensemble of other types of classifiers, demonstrating that the principle of distillation is also applicable to ensembles.

We show that it is possible to apply the distillation method to approximate an ensemble of deep convolutional neural networks with a single network. We do this for two reasons. First, a distilled ensemble will be more portable than the original ensemble, with smaller computational and memory requirements compared to the original version. Second, the improved generalisation achieved with the ensemble may be transferable to the distilled version.

We provide guidelines that we derived from learning theory, as to what the optimal methodology for such a distillation is, and show that, where the original training data is sufficiently sized, the distillation process also serves as a regularizer of the ensemble as a whole. Our recipe for distillation of ensembles also simplifies some of the prerequisites imposed by existing methods: it is no longer necessary to use the "soft" output as a target for training the distilled network. Instead we can use the classification output as the target and the distillation process still works. Additionally, we reason that any additional artificial data used to cover the input space is only likely to reduce the regularization abilities of the distilled network.

The rest of the paper is structured as follows. Section 2 explores the background and existing literature in deep learning, specifically regarding the methods used in this paper. Section 3 covers the necessary background in ensemble theory and methods related to this work. Section 4 presents background on the distillation process. Section 5 derives and illustrates our recipe for the distillation of deep learning ensembles, and Sect. 6 shows how our methodology leads to an improvements over using just the ensemble method. Finally, Sect. 7 discusses our results, and considers how the use of an intermediate ensemble could become an integral part of training a deep neural network.

2 Supervised Deep Learning

Deep Learning is a branch of machine learning applied to the study of very large Artificial Neural Networks (ANNs). Such ANNs are usually termed "deep" because of the large number of hidden layers that they contain. In this paper, we focus our efforts on Convolutional Neural Networks because of their popularity, but our recipe and experimentation can be easily extended to work with other supervised deep learning algorithms.

2.1 Convolutional Neural Networks

Convolutional Neural Networks [11] were introduced mostly for use in image recognition, or other similar problem domains, where applying a convolution operation to the input features makes sense (Video Analysis, Sound data, Natural Language Processing). These networks are usually trained using backpropagation and gradient descent, and most of the readily available update rules can be applied. Each convolutional layer is composed of the following functional elements, which for simplification of the backpropagation steps, can be considered as separate layers that usually come in sequence:

- Convolution of the input features across multiple kernels, sometimes also called filters, where the function of each kernel is learned. In some cases this is thought of as generating a volume of neurons by extrusion, because each kernel will perform a convolution across the input volume. However, the same kernel will be applied on each of the "depth" layers, which leads to weight-sharing and therefore improves the time required to compute and train the layer. If we consider an $N \times M$ rectangular input from layer $l-1$, to which a $n \times m$ rectangular kernel ω is applied, the forward pass of the convolution operation to yield the value of the output at index i, j would be:

$$z_{ij}^{(l)} = \sum_{a=0}^{n-1} \sum_{b=0}^{m-1} \omega_{ab} y_{(i+a)(j+b)}^{(l-1)} \tag{1}$$

An activation function is applied to calculate each output $y_{ij}^{(l)} = A(z_{ij}^{(l)})$, as with standard neurons. When it comes to the back-propagation of the error present at the output of the convolutional layer $E^{(l)}$, we need to compute the partial derivative with respect to each input $\frac{\partial E^{(l)}}{\partial y_{ij}^{(l-1)}}$. This is done using the chain rule, as per normal back-propagation

$$
\begin{aligned}
\frac{\partial E}{\partial \omega_{ab}} &= \sum_{a=0}^{n-1}\sum_{b=0}^{m-1} \frac{\partial E}{\partial z_{ij}^{(l)}} \frac{\partial z_{ij}^{(l)}}{\partial \omega_{ab}} \\
&= \sum_{a=0}^{n-1}\sum_{b=0}^{m-1} \frac{\partial E}{\partial z_{ij}^{(l)}} y_{(i+a)(j+b)}^{(l-1)} \\
&= \sum_{a=0}^{n-1}\sum_{b=0}^{m-1} \frac{\partial E}{\partial y_{ij}^{(l)}} A'(z_{ij}^{(l)}) y_{(i+a)(j+b)}^{(l-1)}
\end{aligned}
\tag{2}
$$

then summing over all $z_{ij}^{(l)}$ where ω_{ab} appears. To propagate the error backwards to the previous layer, we utilize the chain rule again:

$$
\begin{aligned}
\frac{\partial E}{\partial y_{ij}^{(l-1)}} &= \sum_{a=0}^{n-1}\sum_{b=0}^{m-1} \frac{\partial E}{\partial z_{(i-a)(j-b)}^{(l)}} \frac{\partial z_{(i-a)(j-b)}^{(l)}}{\partial y_{ij}^{(l-1)}} \\
&= \sum_{a=0}^{n-1}\sum_{b=0}^{m-1} \frac{\partial E^{(l)}}{\partial z_{(i-a)(i-b)}^{(l)}} \omega_{ab}
\end{aligned}
\tag{3}
$$

- Pooling, or Subsampling, is then applied to the output of the convolution, to reduce the dimensionality of the output. Most commonly the mean or the max functions are applied. If the original back-propagated error on the l-th layer is

$$
E^{(l)} = ((W^{(l)})^T E^{(l+1)}) \cdot A'(z^{(l)}) \tag{4}
$$

  where $E^{(l)}$ is the error term for the l-th layer, W is the weights matrix and $A'(z^{(l)})$ is the derivative of the activation function, we need to first reverse the subsampling, by applying the reverse operation. For instance, for the mean we would just propagate back the error to all inputs to the subsampling:

$$
E^{(l)} = upsample((W^{(l)})^T E^{(l+1)}) \cdot A'(z^{(l)}) \tag{5}
$$

  In practice, this also means that the error propagated backwards from a max-pooling layer is very sparse, because only one of the inputs will receive the error.

Typically, after a number of these convolutional layers, each connected into each other, a number of *fully connected* layers is added, with the specific goal of classifying the input features detected by the convolutional layers.

The "all convolutional neural network" [20], a variant of Convolutional Neural Networks that replaces the pooling layer for dimensionality reduction with a 1×1 convolution, and removes the fully connected layers, has shown improved results on the original method.

3 Ensembles

3.1 Bagging

Bagging (short for "bootstrap aggregating") is a technique that is based on the statistical bootstrapping method, originally introduced in [3], where the original author also shows a number of applied use cases. A quantity N of bootstraps is created by randomly picking M elements from a training dataset of size Z with re-sampling, and then using each of these bootstraps to train a separate identical base classifier. Ref. [3] introduces Bagging with $M = Z$, and this practice seems to be observed in most of the literature. This will create diverse members because of the randomized re-sampling, but because there will be significant overlap in the training sets, all the members will still have positive correlation. If we consider a learner function $Y = \phi(X, T, c)$ that produces a candidate solution Y on class c for a training set T and input vector X, then the aggregate y_j of these bootstraps, by averaging, would be:

$$y_j(c_j) = \frac{\sum_{n=1}^{N} \phi(X, T_n, c_j)}{N} \tag{6}$$

and the final aggregate label is

$$Y = \underset{c_j \in C}{\operatorname{argmax}}\, y_j(c_j) \tag{7}$$

The reason why this works can be argued as follows, and is originally explained, in Ref. [3]. This argument is initially made for regression problems with a follow-up on classification. Given (x, y) pairs taken from the bootstrapped training set T_n, and $\phi(x, T_n)$ as the learned predictor, the aggregate predictor will be the average over all bootstraps:

$$\phi_A(x) = E_T\phi(x, T) \tag{8}$$

If we take x to be a fixed input value and y an output, then

$$E_T(y - \phi(x, T))^2 = y^2 - 2yE_T\phi(x, T) + E_T\phi^2(x, T) \tag{9}$$

Integrating both sides over the joint (x, y) distribution, we get that the MSE of $\phi_A(x)$ is lower than the MSE averaged over T of $\phi(x, T_n)$. The magnitude of this difference is dependent on the difference between the two sides of:

$$(E_T\phi(x, T))^2 \leqslant E_T\phi^2(x, T) \tag{10}$$

Breiman [3] goes on to highlight the importance of the instability between these two figures, and that there is a crossover point at which the bagged example has worse performance.

3.2 AdaBoost

Boosting is a technique first introduced in [18, 19], by which classifiers are trained sequentially, using a sample from the original dataset, with the prediction error from the previous algorithms affecting the sampling weight for the next round. After each round of boosting, the decision can be made to terminate and use a set of calculated weights to apply as a linear combination of the newly created set of learners.

Algorithm 1 A generic Boosting meta-algorithm

$D_0 \leftarrow uniform()$
$t \leftarrow 0$
while *notstoppingConditionReached*() **do**
  $currentSubset \leftarrow pickFromSet(wholeSet, D_t)$
  $h_t \leftarrow newClassifier(currentSubset)$
  $\epsilon_t \leftarrow getError(h_t, wholeSet)$
  $\alpha_t \leftarrow learnerCoefficient(\epsilon_t)$
  $D_{t+1} \leftarrow nextDistribution(D_t, h_t, \alpha_t, wholeSet)$
  $t \leftarrow t + 1$
end while
$H \leftarrow aggregate(h_{0..t}, \alpha_{0..t})$

In [19], Freund and Schapire present two variants of boosting, called AdaBoost. M1 and AdaBoost.M2. The main difference between the two algorithms is in the way the final hypothesis is calculated and how multiple class problems are handled, and they both follow roughly the high-level description of Algorithm 1. Each boosting variant builds a distribution of weights D_t, which is used to sample from the training set, and is updated at each iteration to increase the *importance* of the examples that are harder to classify correctly. The resampled dataset is used to train a new classifier h_t, which is then incorporated in the group, with a weight α_t, based on its classification error ϵ_t. The new D_t is then generated for the next iteration. The main differences between each AdaBoost variant are in how the *getError*, *learnerCoefficient*, *nextDistribution* and *aggregate* functions are implemented.

It has been shown that the requirement of AdaBoost.M1 on maximum error of the underlying weak learner is much stronger than just performing better than random [19]:

The main disadvantage of AdaBoost.M1 is that it is unable to handle weak hypotheses with error greater than 1/2. The expected error of a hypothesis which randomly guesses the label is $1 - 1/k$, where k is the number of possible labels. Thus AdaBoost.M1 requirement for k = 2 is that the prediction is just slightly better than random guessing. However, when $k > 2$, the requirement of AdaBoost.M1 is much stronger than that, and might be hard to meet.

AdaBoost.M2 solves this problem by introducing additional communication between the weak learner and the boosting algorithm. For each sample x and label y a hypothesis function $h_t(x, y) \rightarrow [0, 1]$ is obtained such that the *pseudo-loss* of a hypothesis can be calculated. We use y to indicate the mislabelled class, and y_i to indicate the currently considered weak learner's output associated with x_i. D_t is the weight distribution on the training examples at iteration t.

$$\epsilon_t = \frac{1}{2} \sum_{(i,y)\in B} D_t(i, y)(1 - h_t(x_i, y_i) + h_t(x_i, y)) \tag{11}$$

to be used in place of the error, where B is the set of all mislabelled pairs (i, y):

$$B = (i, y) : i \in 1, \ldots, m, y \neq y_i \tag{12}$$

where m is the number of samples and y_i is the label for sample i. It should be noted that some slight modifications will need to be made to standard learners in order to produce this hypothesis, so that instead of producing a single decision, they can output a probability $P(y|h_t, x) = h_t(x, y)$ of y being the correct label assignment.[1] Subsequently, we set $\beta_t = \epsilon_t/(1 - \epsilon_t)$ and use it to calculate the updated value of D_t:

$$D_{t+1}(i, y) = \frac{D_t(i, y)}{Z_t} \cdot \beta_t^{\frac{1}{2}(1+h_t(x_i,y_i)-h_t(x_i,y))} \tag{13}$$

where Z_t is the normalization constant that is also calculated in AdaBoost.M1.

3.3 Deep Incremental Boosting

Deep Incremental Boosting (DIB), introduced in [14], is an example of a *white-box* ensemble method: instead of treating the base learner as a "black-box", characteristics and properties of such model are exploited to improve the results given by the ensemble method. DIB applies concepts from transfer of learning to improve the speed of convergence of boosting rounds by copying a subset of convolutional layers

[1]Fortunately, in the case of Deep Learning algorithms, we find that in most cases the output layer is a SoftMax, which already gives us this value with no additional work.

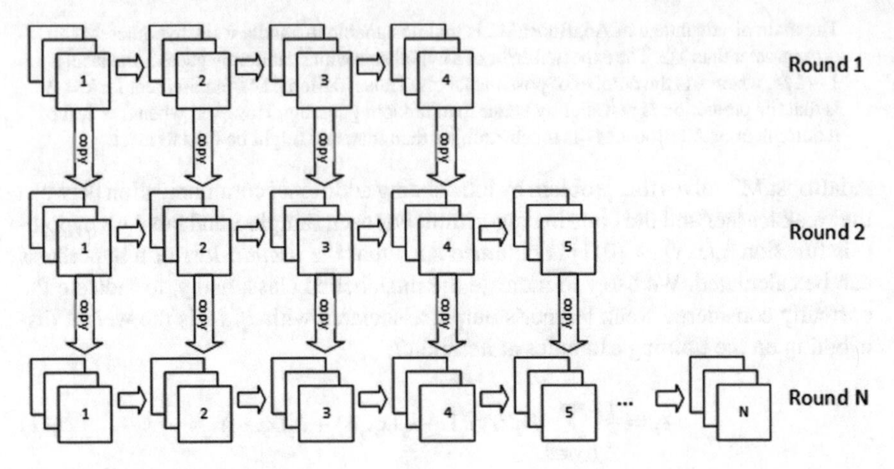

Fig. 1 Illusration of subsequent rounds of DIB

trained at round $t-1$ as the initialisation for the network to be trained at round t. The new network is also given additional capacity to learn the new resampled dataset and the corrections from the previously trained layers. An illustration of the architecture of DIB is given in Fig. 1, and the full algorithm is shown in Algorithm 2.

Algorithm 2 Deep Incremental Boosting

$D_0(i) = 1/M$ for all i
$t = 0$
$W_0 \leftarrow$ randomly initialised weights for first classifier
while $t < T$ **do**
  $X_t \leftarrow$ pick from original training set with distribution D_t
  $u_t \leftarrow$ create untrained classifier with additional layer of shape L_{new}
  copy weights from W_t into the bottom layers of u_t
  $h_t \leftarrow$ train u_t classifier on current subset
  $W_{t+1} \leftarrow$ all weights from h_t
  $\epsilon_t = \frac{1}{2}\sum_{(i,y)\in B} D_t(i)(1 - h_t(x_i, y_i) + h_t(x_i, y))$
  $\beta_t = \epsilon_t/(1 - \epsilon_t)$
  $D_{t+1}(i) = \frac{D_t(i)}{Z_t} \cdot \beta^{(1/2)(1+h_t(x_i,y_i)-h_t(x_i,y))}$
  where Z_t is a normalisation factor such that D_{t+1} is a distribution
  $\alpha_t = \frac{1}{\beta_t}$
  $t = t + 1$
end while
$H(x) = \underset{y\in Y}{\operatorname{argmax}} \sum_{t=1}^{T} log\alpha_t h_t(x, y)$

---

4 Related Work

4.1 Meta–Models

In the field of simulation, the concept of a *meta–model* is not a novel one. A meta–model is a model that is designed to reproduce the effects of another model.

This "second–order" modelling has several advantages. It may be the case that the original model cannot be used to make predictions, or it may be too complex to fully evaulate, so a meta–model can be developed to obtain similar results with lesser cost. It may be that some of the characteristics of the original model are not measurable, so that it cannot be completely reproduced.

A number of surveys has been produced on the subject of meta–modelling, for example [22]. The meta–models that are more closely related to the work in this paper are those that use Artificial Neural Networks as the second–order model [2, 7]. These meta–models typically have been trained on small datasets and the ensemble member is also usually a model with relatively small capacity. The distilled ANN cannot be considered deep enough to be representative of modern deep learning.

4.2 Approximating an Ensemble with a Neural Network

In [23] it has been shown that it is possible to use a neural network to approximate an ensemble. However, whilst the results from the approximate network are better than training the network from scratch, and they approximate very well the results produced by utilising bagging, the ANNs used are also shallow and very small compared to today's deep learning models. This method makes use of a *pseudo-training set* which is generated from the original training set and the outputs of the ensemble.

An improvement to this methodology has been developed by [4]. Instead of using a sample of the original training set to generate the pseudo-training set, the authors sample the entire input space S. This approach is then evaluated on trying to approximate bagged ensembles of SVMs and numerous variants of decisions trees. The generation of artificial data points in the sample space appears to work well for small datasets with low complexity. However, as we will see, for high-dimensional datasets such as those used in deep learning, it poses a few problems of its own.

4.3 Distillation in Deep Learning

Work has been done towards showing that it is possible to distil the knowledge in an ensemble of relatively deep neural networks [9]. In the paper presenting this work, a similar argument as in [23] is made: the true function underpinning the data is not known and therefore it is impossible to learn to generalise it perfectly using only what is represented by the training data, so a *proxy* function has to be used—in this case the output of the cumbersome model. This gives credit to the idea of generating a pseudo-training dataset to extract the knowledge of said proxy function, which in this case is to be created using the inputs of the original training set (or a specially dedicated "transfer set") and the *soft* targets for those given inputs. For a single deep neural network, this equates to the output of the Softmax layer, before any argmax operation is applied. In the case of ensembles, this is extended as the geometric or algebraic mean of the soft outputs of each ensemble member. It is argued that doing so serves as a regulariser.

This research is focused on relatively small deep networks (up to 8 layers), and without any of the additional architectural characteristics that are typically found in contemporary deep learning systems. Additionally, two of the experiments provided are not reproducible, as they are conducted on ad-hoc, unpublished, proprietary datasets (Google JFT and unnamed Android development set). On MNIST, the only results given are for a small network, used to distill a larger network. The large network reports an error of 0.67%, the small one 1.46% and the distilled version 0.74%, indicating that the process of distillation is useful to improve the performance of the smaller network, but it does not beat the performance of the cumbersome one. An attempt is then made to utilise ensembles in the distillation process. However, the authors make use of "specialist models", where each member of the ensemble has received specialised training on a specific class or task, and no consideration is made for the general case of traditional ensemble methods.

We therefore conclude that, although it is seminal work in proving that distillation of deep learning models is possible, it is still necessary to extend the experimentation to more modern supervised deep learning architectures, to network sizes that are similar to those currently being used, and, most importantly, to non-specialised ensembles.

Additional work has been done when looking at distillation as a tool, especially with regards to *adversarial examples* [8, 16, 17]. Such examples are generated in a way that a small perturbation of the input, which is normally imperceptible to the human eye, causes a large change in the outputs of a network. It is shown that distillation reduces the magnitude and count of input gradients that create these adversarial examples, which serves as a good indication that the process of distillation is very effective at removing the superfluous perturbations of the learned function, by simplifying it and regularising it.

Another variant of distillation has been developed, called FitNet [1], which specifically trains a distilled network that has a larger number of smaller layers than its teacher. In this case, we can see that a teacher network that has 9.84% error on CIFAR-10, can be used to train a FitNet network that reaches 8.39% error on the same test set.

5 Distillation of Deep Learning Ensembles

We begin by formally defining the process of distillation.

Definition 1 Let S be the (unknown) manifold of the feature space $\mathbb{S}$ where some data X exists.

Let Y be the set of labels corresponding to the data X.

Let $\mathbb{Y}$ be the output space in which the data is labelled.

Let $Y = f(X) : \mathbb{S} \rightarrow \mathbb{Y}$ be the (unknown) *ground truth* function that correctly labels the data, which is the target of the learning.

Let D be the (unknown) distribution of the data X in the manifold S.

Let $\bar{D}$ be an approximate estimation of the distribution D.

Let $h(x)$ be a model hypothesis that is designed to learn the function $f(x)$.

Let $\bar{Y} = \bar{f}(X) : \mathbb{S} \rightarrow \mathbb{Y}$ be the *approximate* function that is instead learned by the model $h(X)$.

Let $\bar{X}$ be any additional synthetic training data sampled from $\bar{D}$ and $\bar{f}(\bar{X})$ its predicted labels by the model $h(X)$.

Definition 2 Then we establish that it is possible to derive a *distilled* model $h'(X)$ which learns from data $\left(\bar{X}, \bar{f}(\bar{X})\right)$ a new function $\hat{f}(X) : \mathbb{S} \rightarrow \mathbb{Y}$ which is itself an approximation of $\bar{f}(X)$.

Figure 2 represents graphically how each function is learned from its ancestor, mapping the same two spaces in different manners. Because the functions $f(X), \bar{f}(X)$ and $\hat{f}(X)$ are by definition increased orders of approximation of one another, it is not possible to assume that they all cover the same manifold of the spaces $\mathbb{S}$ and $\mathbb{Y}$. We therefore identify the manifolds as:

- S and Y for $f(X)$
- $\bar{S}$ and $\bar{Y}$ for $\bar{f}(X)$
- $\hat{S}$ and $\hat{Y}$ for $\hat{f}(X)$

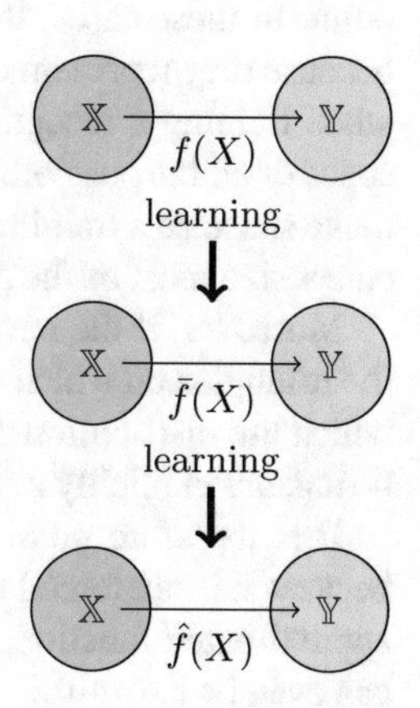

Fig. 2 A schematic representation of distillation

Proposition 1 *X is the best available representation of the feature space manifold S at training time.*

Because the distribution D is unknown, any estimation of it will be an approximation and inevitably

$$\bar{D} - D \neq \varnothing \tag{14}$$
$$\bar{D} - D \notin S \tag{15}$$

This implies that any new synthetic example $\bar{x}$ may be possibly drawn from $\bar{D} - D$, and that therefore there is a non-zero probability $P(\bar{x} \notin S) > 0$ of $\bar{x}$ being sampled outside of the manifold S.

Proposition 2 *Any additional training example $\bar{x}$ sampled from $\bar{D} - D$ does not provide any additional information to improve the generalisation from $\hat{f}(X) \to f(X)$.*

We know that $\bar{f}(X) \neq f(X)$, because otherwise the learning for the particular problem would be solved. It follows that any individual samples $\bar{x}$ taken outside of S will have no use for the representation of $f(X)$ when training $\hat{f}(X)$, because these regions are outside the domain of $f(X)$.

Proposition 3 *As the manifold S becomes smaller with respect to the feature space $\mathbb{S}$, the data is more sparse, and therefore $P(\bar{x} \notin S)$ increases.*

Paradoxically, for an infinitely small manifold S, we know that all the synthetic data will lie outside of S, and therefore represent values of $\bar{f}(X)$ that are not defined in $f(X)$. Therefore, although the synthetic dataset $\bar{X}$ will improve the ability of generalising $\bar{f}(X)$ from $\hat{f}(X)$, it will have a higher likelihood of containing information with a higher deviation from the original target function $f(X)$.

In the case of common problems in Deep Learning, we know S to be very small compared to $\mathbb{S}$, therefore making the additional purely synthetic data superfluous, and in some cases potentially harmful to the learning. This is in contrast with noise–injection and data–augmentation, which use existing training examples as a starting point. In these cases, the augmented examples can also be considered as part of S because they represent classifiable data. In practical terms, it is easy to imagine that, when training a recognizer of hand-written digits, not all the values of the feature space of all the possible pixel value combinations in an image of the same dimension make sense as a training example. In fact, it makes no sense semantically to try and categorize most of the possible resulting images as a digit (Fig. 3).

Moreover, if the new training set for the small network is generated solely from the examples on which the Ensemble is trained, other areas of the feature space $\mathbb{S}$ in which the distribution D does not exist, and the approximate function $\bar{f}(X)$ is overfitting, are explicitly avoided.

It is therefore possible to consider the smaller model as a regularised model, because it is not forced to learn these overfitted regions, allowing it to learn a *simpler* approximated function. Because of this, any additional sample $\bar{x}$ is superfluous and can even be harmful.

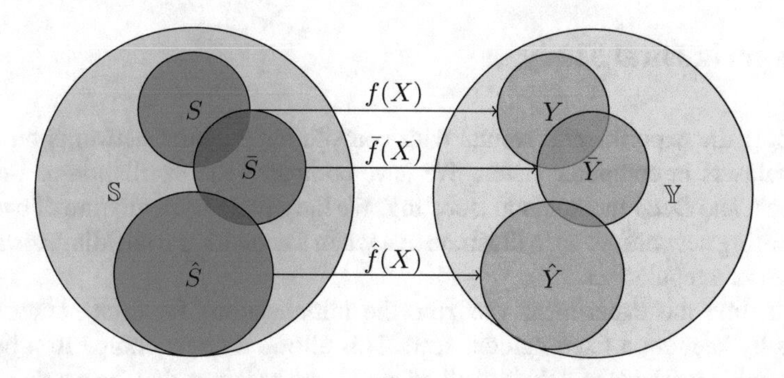

Fig. 3 A schematic depiction of how manifolds might be overlapping

Because the given definition of distillation uses $h(X)$ as the target output, the proposed approach does not make use of soft target probabilities and does not create additional synthetic training data. This recipe relies solely on the assumption that the original Ensemble is likely to have overfit the training data. The soft target probabilities are not used as simplification to the procedure. We found empirically that with ensembles of deep convolutional neural networks the difference was small enough to justify dropping this practice in favour of simplifying the methodology.

By obtaining the labels produced by the Ensemble on the training set X_{train} and using them to train a new classifier of the same type and shape as those used as members of the Ensemble (Algorithm 3), it is possible to construct a regularised version of the approximated function $\bar{f}(X)$, which is learned by a model $h(X)$.

The validation and testing of the model is performed on the original validation and test sets. This ensures that the distilled model's learned function $\hat{f}(X)$ is evaluated on how well it approximates $f(X)$—the (unknown) function that correctly labels the data, which is the target of the learning—rather than $\bar{f}(X)$, which was the original goal of the learning process.

Algorithm 3 Regularized Distillation

$H(X,Y) =$ generator of deep networks
$E(H,X,Y) =$ ensemble method
$h_E(X,Y) \leftarrow E(H, X_{train}, Y_{train})$
$y_h(X_{train}) \leftarrow h_E(X_{train}, Y_{train})$
$h_D(X, y_h(X)) \leftarrow H(X_{train}, y_h(X)$
evaluate $h_D(X, y_h(X))$ on the original validation and testing sets (X_{valid}, Y_{valid}) and (X_{test}, Y_{test})

Although in theory it is possible to utilise any architecture for the distilled network, it is practical to use the same architecture originally used for the base classifier. This avoids the need to fit additional hyperparameters and the architecture is already known to be suitable to the problem. It also serves to demonstrate the point that improved learning can be achieved with no additional capacity.

6 Experimental Study

We report the experimental results with convolutional neural networks on benchmark datasets in computer vision. We have compared the distillation of Bagging, AdaBoost and Deep Incremental Boosting. We have included a mixture of over- and under-fitting network, so as to illustrate that when a network is overfitting, distillation also acts as a regulariser.

Each time the experiment was run, the initialisations for each network were aligned by keeping a fixed random seed. This allows a direct comparison between the ensemble methods and their distilled counterparts, given that the single network is initialised identically each time. The values reported are for the median result from five separate runs. The goal of this experimental study is not to achieve a new state–of–the–art performance on the datasets under examination, but to demonstrate how our methodology can be applied to improve a well–performing network, that is close to state–of–the–art.

All experiments were run with the Toupee Deep Learning Ensemble experimentation toolkit. This is a set of libraries and tools, based on Keras [5], which allow the creation of repeatable experiments in deep learning, including ensemble methods. An experiment is described by a "model file", which contains the entire description of the architecture of the network, and an "experiment file", which contains all the information needed to train and test the network, and if applicable, the ensemble. Source code for Toupee is available at http://github.com/nitbix/toupee.

6.1 MNIST

MNIST [12] is a common computer vision dataset that associates pre-processed images of hand-written numerical digits with a class label representing that digit. The input features are the raw pixel values for the 28×28 images, in grayscale, and the outputs are the numerical value between 0 and 9.

MNIST is generally considered a solved problem, with some relatively simple deep learning models reaching 99.79% accuracy of the best model on the test set with the appropriate data augmentation [21] and longer training. However, we have used this dataset to show an example of how an ensemble of relatively small models that has already slightly overfit the data can be approximated by our regularised distillation. Figure 4 shows how even such a simple convolutional network is already able to start overfitting the training set, by reaching 0% classification error on the training set within four epochs, whilst the test error starts increasing after epoch 14.

We note from Table 1 that the distilled classification errors are better than the ensemble for each of the methods considered, and also better than the original single network (0.66%), further corroborating our hypothesis that the distillation process can serve as a regulariser.

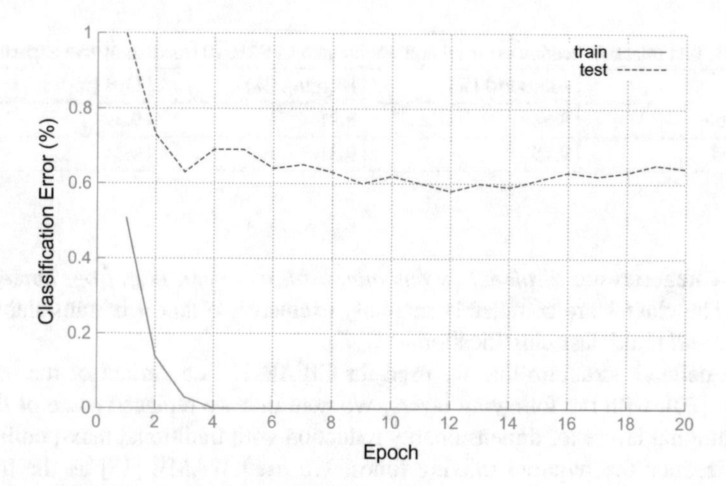

Fig. 4 Non-augmented training and test error on MNIST

Table 1 Test misclassification error on MNIST (median of five experiments)

	AdaBoost (%)	Bagging (%)	DIB (%)
Ensemble	0.63	0.59	0.51
Distilled	0.52	0.55	0.49

The Convolutional Neural Network–CNN used for MNIST has the following structure. We used WAME [15] as the training method, and trained for 20 epochs. We found that training on MNIST for additional time did not improve generalization. We used the training, validation and test sets provided, and commonly used in the literature.

- An input layer of 784 nodes
- 64 5×5 convolutions
- 2×2 max-pooling
- 128 5×5 convolutions
- 2×2 max-pooling
- A fully connected layer of 1024 nodes
- Dropout with $P(D) = 0.5$
- a Softmax layer with 10 outputs (one for each class).

6.2 CIFAR-10

CIFAR-10 is a dataset that contains 60,000 small images of 10 categories of objects. It was first introduced in [10]. The images are 32×32 pixels, in RGB format. The

Table 2 Test misclassification error on non-augmented CIFAR-10 (median of five experiments)

	AdaBoost (%)	Bagging (%)	DIB (%)
Ensemble	9.62	8.91	9.39
Distilled	9.15	9.31	9.34

output categories are *airplane, automobile, bird, cat, deer, dog, frog, horse, ship, truck*. The classes are completely mutually exclusive so that it is translatable to a *1-versus-all* multiclass classification.

The network structure that we used for CIFAR-10 is a variant of the network used in [20], with the following layers. We note that we replaced some of the all-convolutional layers for dimensionality reduction with traditional max-pooling layers, to reduce the required training times. We used WAME [15] as the training method. The training lasted 100 epochs, and, as is practice with the CIFAR datasets, did not use a validation set.

- An input layer of $3 \times 32 \times 32$ nodes
- 96 3×3 convolutions, with 1×1 padding
- 96 3×3 convolutions, with 1×1 padding
- 2×2 max-pooling
- Dropout with $P(D) = 0.5$
- 192 3×3 convolutions, with 1×1 padding
- 192 3×3 convolutions, with 1×1 padding
- 2×2 max-pooling
- Dropout with $P(D) = 0.5$
- 192 3×3 convolutions, with 1×1 padding
- 192 3×3 convolutions, with 1×1 padding
- 192 1×1 convolutions, with 1×1 padding
- 10 1×1 convolutions, with 1×1 padding
- Dropout with $P(D) = 0.5$
- Global average pooling
- a Softmax layer with 10 outputs (one for each class).

We first trained our model with no dataset augmentation, running the experiment five times. The single network obtained a median classification error of 11.15%. This is higher than the original all-convolutional network, which is reported as 9.08%. The principal reasons for the difference are:

- our dataset was not preprocessed with ZCA whitening and Global Contrast Normalization.
- the shorter and more aggressive training schedule, that we used in order to be able to run ensembles of 10 models in an acceptable time.
- the original paper is not reporting the median of five runs, but appears to just be reporting the best results.

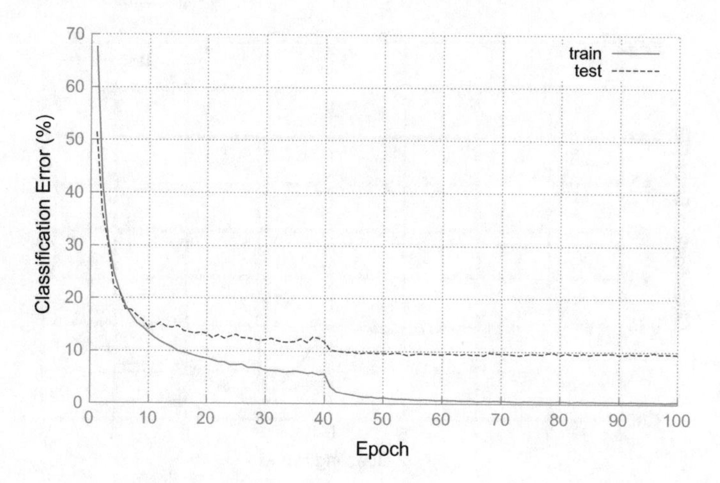

Fig. 5 Non-augmented training and test error on CIFAR-10

Table 3 Test misclassification error on augmented CIFAR-10 (median of five experiments)

	AdaBoost (%)	Bagging (%)	DIB (%)
Ensemble	7.64	6.69	7.16
Distilled	7.14	8.14	7.11

We were able to recreate a result that was similar to the original reported values when using the full training schedule, but given the long time required to train we were not able to incorporate this into the ensembles. This however, means that the network is underfitting the data, and therefore it is also likely that the ensemble will be doing the same. This is corroborated by the graph in Fig. 5, which shows that even at the final epochs of training there is still improvement in the test error.

Each ensemble method was run with 10 members, then the final resulting ensemble was distilled according to the recipe provided in Sect. 5. The results without any data augmentation are reported in Table 2. We then repeated the same experiment with data augmentation, as prescribed in [13]. Results for the augmented data are reported in Table 3. In this case our single network reaches a median error of 10.07%.

We believe that, while Bagging might still be able to improve the generalisation and is therefore slightly underfitting (despite achieving the best results of the three ensemble methods), Deep Incremental Boosting and AdaBoost are able to overfit the data, because of the increase in learning capacity added at each round and additional emphasis on "hard-to-classify" examples. This is corroborated by the graph in Fig. 6, which shows how subsequent rounds of DIB are not improving the performance of the ensemble on the test set.

We note how the process of distillation, once again, has been able to improve the generalisation results for those ensembles that were overfitting, while, although

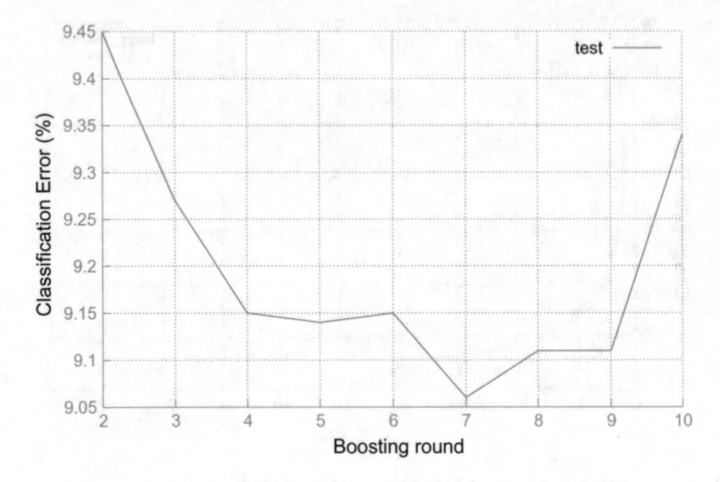

Fig. 6 Rounds of DIB on augmented CIFAR-10 show overfitting

it wasn't able to improve the results for bagging, it is still returning considerably improved results when compared to the single network.

6.3 CIFAR-100

CIFAR-100 is a dataset that contains 60000 small images of 100 categories of objects, grouped in 20 super-classes. It was first introduced in [10]. The image format is the same as CIFAR-10. Class labels are provided for the 100 classes as well as the 20 super-classes. A super-class is a category that includes 5 of the fine-grained class labels (e.g. "insects" contains *bee, beetle, butterfly, caterpillar, cockroach*).

The network used was identical to that used to classify the CIFAR-10 dataset, except for the number of filters in the last 1×1 convolution and the number of outputs of the Softmax, which were both increased to 100 to reflect the change in the number of output classes. We also used the same training parameters.

We first trained our model with no dataset augmentation. The single network obtained a median classification error of 39.48%. This is again higher than the original all-convolutional network, which is reported as 33.71%. We believe that the reason for not obtaining the same error results as reported in the original paper are the same as those we reported for CIFAR-10. Additionally, the shorter training schedule allowed for a slight underfitting of the single network. This is corroborated by the graph in Fig. 7, which shows that even at the final epochs of training there is still improvement in the test error.

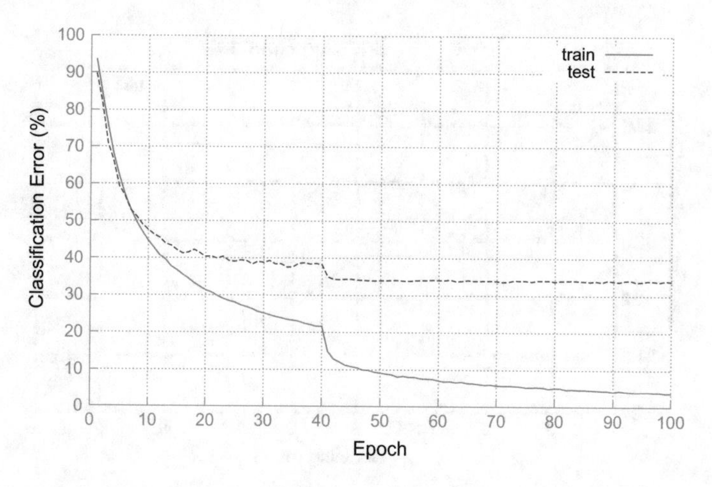

Fig. 7 Non-augmented training and test error on CIFAR-100

Table 4 Test misclassification error on non-augmented CIFAR-100 (median of five experiments)

	AdaBoost (%)	Bagging (%)	DIB (%)
Ensemble	32.23	31.41	31.40
Distilled	33.66	34.12	33.69

Table 5 Test misclassification error on augmented CIFAR-100 (median of five experiments)

	AdaBoost (%)	Bagging (%)	DIB (%)
Ensemble	31.49	28.58	30.38
Distilled	32.56	31.02	31.32

Each ensemble method was run for 10 rounds, then the final resulting ensemble was distilled according to the recipe provided in Sect. 5. The results without any data augmentation are reported in Table 4. We then repeated the same experiment with data augmentation, as prescribed in [13]. Results for the augmented data are reported in Table 5. The single network with augmentation reaches an error of 32.76%.

We believe that, unlike with CIFAR-10, the ensemble methods have not been able to overfit the data. This is corroborated by the graph in Fig. 8, which shows how subsequent rounds of DIB are improving the performance of the ensemble on the test set, suggesting that if there were further rounds, the performance would continue to improve.

We note how, even though we are not improving on the results of the ensemble, the distillation process does improve the results when compared to the single network, capturing almost all of the gains given by the ensemble.

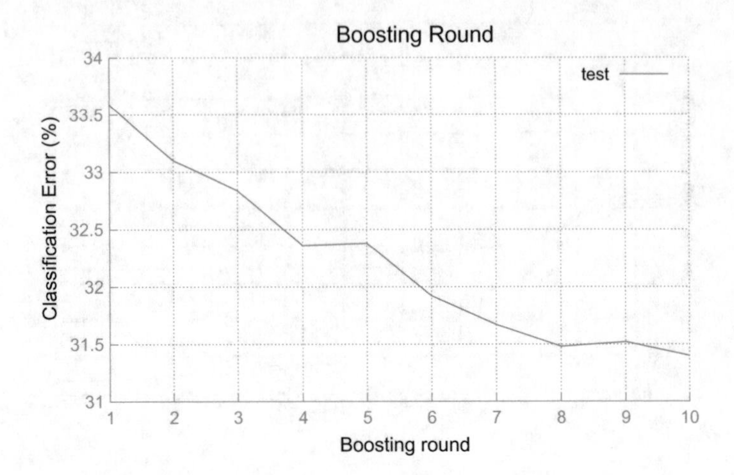

Fig. 8 Rounds of DIB on augmented CIFAR-100 show no overfitting

We then repeated the same experiment with dataset augmentation: random horizontal flips, random zoom and tilt, but no random cropping. We still did not use ZCA whitening or Global Contrast Normalisation. The results in Table 5 show that a similar conclusion to the non-augmented version of the experiment can be made: because the Ensemble is not overfitting, the distillation process is regularising a function that doesn't need to be regularised. We note that, although the distilled network has worse performance than the ensemble methods, it still has better performance than the original network (33.71%).

7 Discussion and Conclusions

We have seen how the process of distillation of a cumbersome model can be extended to the representation of the function learned by an ensemble of deep neural networks with a single deep neural network. We have reported some guidelines that, when appropriate, also show how this process can be utilised as a regulariser for the function learned by the ensemble.

We have explored experimentally how our methodology works, both on overfitted and underfitted networks, showing that in all cases we are able to improve the performance of the single network by distilling the knowledge of the ensemble. Furthermore, when the ensemble is overfitting the training data, we are able to use the distillation process as a regulariser, improving the generalisation.

Based on the data from all experiments, a single-tailed Wilcoxon test confirms that, at the 5% confidence level, the improvements in performance are significant. The observed empirical evidence when comparing the distilled network to the original single network shows that the distilled network has lower error every time. A single-tailed Wilcoxon test also confirms that the distilled network is significantly better at the 5% confidence level. Because of the results of these two tests, we can conclude that, in the cases where the computational capacity is not sufficient to run the ensemble at test time (for example those situations where the model needs to be run at scale on commodity hardware), it is always better to use the distilled network.

In terms of the regularisation ability of the distillation process, our experiments provide evidence that distillation works best in cases where the original ensemble is overfitting the data. In such situations, applying distillation improves the accuracy even further. If we imagine that a single member network can learn a function of complexity C_n, we can assume that the overall learning capacity for an ensemble of size n will be $C_e \approx nC_n$. The process of distillation subsequently reduces the complexity of the function learned back to C_n. However, because of the changed learning process explained in Sect. 5, the function being learned is more effective and is able to reduce the overfitting learned by the ensemble.

Lastly, another finding of this work is that utilising a distilled ensemble, even in situations where the distillation does not provide good regularisation, can still produce improvements over the original single network. We therefore can easily envision situations where training an ensemble and distilling its knowledge back to a single network could become an integral part of training the single network. In these cases, the single network remains the goal of the training process, and the ensemble becomes a training vehicle for the improvement of the generalisation.

References

1. Adriana, R., et al.: Fitnets: hints for thin deep nets. arXiv preprint arXiv:1412.6550 (2014)
2. Badiru, A.B., Sieger, D.B.: Neural network as a simulation metamodel in economic analysis of risky projects. Eur. J. Oper. Res. **105**(1), 130–142 (1998)
3. Breiman, L.: Bagging predictors. Mach. Learn. **24**(2), 123–140 (1996)
4. Bucilu, C., Caruana, R., Niculescu-Mizil, A.: Model compression. In: Proceedings of the 12th ACM SIGKDD International Conference on Knowledge Discovery and Data Mining, pp. 535–541. ACM (2006)
5. Chollet, F.: keras. https://github.com/fchollet/keras (2015)
6. Dietterich, T.: Ensemble methods in machine learning. In: Multiple Classifier Systems. Lecture Notes in Computer Science, vol. 1857, pp. 1–15. Springer, Berlin (2000)
7. Fonseca, D., Navaresse, D., Moynihan, G.: Simulation metamodeling through artificial neural networks. Eng. Appl. Artif. Intell. **16**(3), 177–183 (2003)
8. Goodfellow, I.J., Shlens, J., Szegedy, C.: Explaining and harnessing adversarial examples. arXiv preprint arXiv:1412.6572 (2014)
9. Hinton, G., Vinyals, O., Dean, J.: Distilling the knowledge in a neural network. arXiv preprint arXiv:1503.02531 (2015)
10. Krizhevsky, A., Hinton, G.: Learning Multiple Layers of Features from Tiny Images (2009)
11. LeCun, Y., Bengio, Y.: Convolutional networks for images, speech, and time series. In: The Handbook of Brain Theory and Neural Networks, vol. 3361, p. 310 (1995)
12. Lecun, Y., Cortes, C.: The MNIST database of handwritten digits. http://yann.lecun.com/exdb/mnist/
13. Lin, M., Chen, Q., Yan, S.: Network in network. arXiv preprint arXiv:1312.4400 (2013)

14. Mosca, A., Magoulas, G.: Deep incremental boosting. In: Benzmuller, C., Sutcliffe, G., Rojas, R. (eds.) GCAI 2016. 2nd Global Conference on Artificial Intelligence. EPiC Series in Computing, vol. 41, pp. 293–302. EasyChair (2016)
15. Mosca, A., Magoulas, G.D.: Training convolutional networks with weight-wise adaptive learning rates. In: ESANN 2017 Proceedings, European Symposium on Artificial Neural Networks, Computational Intelligence and Machine Learning. Bruges (Belgium), 26–28 April 2017, i6doc.com publ. (2017)
16. Papernot, N., McDaniel, P., Goodfellow, I., Jha, S., Celik, Z.B., Swami, A.: Practical black-box attacks against deep learning systems using adversarial examples. arXiv preprint arXiv:1602.02697 (2016)
17. Papernot, N., McDaniel, P., Wu, X., Jha, S., Swami, A.: Distillation as a defense to adversarial perturbations against deep neural networks. In: 2016 IEEE Symposium on Security and Privacy (SP), pp. 582–597. IEEE (2016)
18. Schapire, R.E.: The strength of weak learnability. Mach. Learn. **5**, 197–227 (1990)
19. Schapire, R.E., Freund, Y.: Experiments with a new boosting algorithm. In: Machine Learning: Proceedings of the Thirteenth International Conference, pp. 148–156 (1996)
20. Springenberg, J.T., Dosovitskiy, A., Brox, T., Riedmiller, M.: Striving for simplicity: The all convolutional net. arXiv preprint arXiv:1412.6806 (2014)
21. Wan, L., Zeiler, M., Zhang, S., Cun, Y.L., Fergus, R.: Regularization of neural networks using dropconnect. In: Proceedings of the 30th International Conference on Machine Learning (ICML-13). pp. 1058–1066 (2013)
22. Wang, G.G., Shan, S.: Review of metamodeling techniques in support of engineering design optimization. J. Mech. Des. **129**(4), 370–380 (2007)
23. Zeng, X., Martinez, T.R.: Using a neural network to approximate an ensemble of classifiers. Neural Process. Lett. **12**(3), 225–237 (2000)

Heuristic Constraint Answer Set Programming for Manufacturing Problems

Erich C. Teppan and Gerhard Friedrich

Abstract Constraint answer set programming (CASP) is a family of hybrid approaches integrating answer set programming (ASP) and constraint programming (CP). These hybrid approaches have already proven to be successful in various domains. In this paper we present the CASP solver ASCASS (A Simple Constraint Answer Set Solver) which provides novel methods for defining and exploiting search heuristics. Beyond the possibility of using already built-in problem-independent heuristics, ASCASS allows on the ASP level the definition of problem-dependent variable selection, value selection and pruning strategies which guide the search of the CP solver. In this context, we investigate the applicability and performance of CASP in general and ASCASS in particular in two important manufacturing problem domains: system configuration and job scheduling.

1 Introduction

During the last decade, Answer Set Programming (ASP) under the stable model semantics [13] has evolved to an extremely powerful approach for solving combinatorial problems. Especially conflict-driven search mechanisms contributed to the high performance of state-of-the-art solvers [11]. Furthermore, ASP provides superior problem encoding capabilities as ASP is declarative in nature and even provides language features which go beyond first order.

However, the expressive power on the one hand and the highly effective conflict-driven search approach on the other hand does not come for free. Current ASP solvers employing conflict-driven search transform the higher-order problem representation to propositional logic. This transformation (called grounding) constitutes the space bottleneck of nowadays ASP systems. Once the grounding step is completed, the per-

E.C. Teppan (✉) · G. Friedrich
Alpen-Adria Universit Klagenfurt, Klagenfurt, Austria
e-mail: erich.teppan@aau.at

G. Friedrich
e-mail: gerhard.friedrich@aau.at

I. Hatzilygeroudis and V. Palade (eds.), *Advances in Hybridization of Intelligent Methods*, Smart Innovation, Systems and Technologies 85,
https://doi.org/10.1007/978-3-319-66790-4_7

formance of the conflict-driven search in combination with state-of-the-art look-back heuristics like VSIDS and restarts [17] typically shows superior performance compared to other search approaches. Yet, grounding is not possible for many industrial-sized problem instances.

One approach that emerged also out of the need of easing the grounding was Constraint Answer Set Programming (CASP) [19]. CASP can be seen as a hybrid approach extending ASP by Constraint Programming (CP) features. Conceptually, it is very close to satisfiability (SAT) modulo theory approaches which integrate first-order formulas with additional background theories such as real numbers or integers [24]. For certain classes of problems like special forms of scheduling CASP was already successfully applied [3]. Especially search problems with large variable domains often profit from the CASP representation due to the alleviation of the grounding bottleneck [18]. In CASP often a majority of the solution calculation is done by the CP solver. Hence, the applied search strategies on the CP level play a crucial role for the successful application of CASP in real-world problem domains. Furthermore, for many real-world problem domains problem-independent general purpose strategies typically built-in state of-the-art solvers are not sufficient and problem-dependent heuristics are needed.

Any problem-dependent heuristic on the CP level basically consists of three components:

1. a problem-dependent variable selection strategy
2. a problem-dependent value selection strategy
3. a problem-dependent pruning strategy

Up to now there was no focus on the development of sophisticated features for expressing and exploiting search strategies in CASP solvers. Consequently, the support for expressing and exploiting search strategies on the CP level are rather limited in existing solvers. The two most well known CASP solvers are Clingcon and EZCSP.

Clingcon[1] [21] does not provide any means for influencing the search of the underlying CP solver Gecode.[2] In EZCSP[3] [2] there is a set of built-in strategies depending on the underlying CP solver that can be used. In EZCSP, also problem-dependent variable selection strategies are already possible. By the special predicate *label_order* it is possible to define the order in which the constraint variables are processed by the CP solver. What is missing is the possibility of expressing custom value ordering and pruning strategies.

In this article we first present ASCASS, a novel CASP solver which uses Clingo for answer set solving and the Java framework Jacop for CP solving. ASCASS combines and extends the heuristic capabilities of state-of-the-art CASP solvers and makes them completely available on the problem encoding level. Beyond the usage of built-in strategies, ASCASS provides powerful constructs for the formulation and

[1]http://www.cs.uni-potsdam.de/clingcon/.

[2]www.gecode.org.

[3]http://www.mbal.tk/ezcsp/index.html.

exploitation of problem-dependent heuristics consisting of variable selection, value selection and pruning strategies.

Based on the heuristic representation abilities of ASCASS, we show how to realize heuristic methods for two industrially highly important problem fields. The first problem field is concerned with the configuration of products and services. As a generic representative for such problems we investigate the partner units configuration problem. The second problem field is job scheduling in manufacturing environments. Here, we focus on the most famous job-shop scheduling problem. We provide a new benchmark of realistic large-scale instances with proven optimal solutions comprising up to 10,000 job operations scheduled on up to 100 machines. In particular, we show how to realize dispatching rules. Dispatching rules are greedy and easy to calculate scheduling rules that are widely applied in nowadays production lines.

The remainder of this article is structured as follows: The next section gives a small introduction into the basics of ASP, CP and CASP. Section 3 introduces ASCASS, a simple constraint answer set solver. Section 4 examines configuration problems as an industrially important problem field and shows how to realize heuristic problem solving of the partner units configuration problem within the CASP framework. Section 5 discusses job scheduling as an important problem class for industry. Based on a new set of benchmark instances with proven optimal solutions, we investigate the application of ASCASS on large-scale job-shop scheduling. In particular, we show how to produce schedules with the help of dispatching rules. Finally, Sect. 6 summarizes the main results and concludes the article.

2 Background

In this section we introduce the basic concepts of answer set and constraint answer set programming as it is needed for the purposes of this article. In particular, we ignore disjunctive logic rules and classic negation in ASP for readability reasons. For further information about ASP and CASP please refer to [2, 11, 13, 19, 21].

2.1 Syntax of ASP

In ASP, a *term* refers either to a (logic) *variable* or a *constant*. Strings starting with upper case letters denote variables. Constants are represented by strings starting with lower case letters, by quoted strings or by integers. An *atom* is either a *classical atom*, a *cardinality atom* or an *aggregate atom*. A classical atom is an expression $p(t_1, \ldots, t_n)$ where p is an n-ary predicate and $t_1, \ldots, t_n$ are terms. A *negation as failure (NAF) literal* is either a classical atom λ or its negation *not* λ. A cardinality literal is either a cardinality atom ψ or its negation *not* ψ. A *cardinality atom* is of the form

$$l \prec_l \{a_1 : l_{1_1}, \ldots, l_{1_m}; \ldots; a_n : l_{n_1}, \ldots, l_{n_o}\} \prec_u u$$

where

- $a_i : l_{i_1}, \ldots, l_{i_j}$ represent *conditional literals* in which a_i (the heads of the cardinality atom) constitute classical atoms and the conditions l_{i_j} are NAF literals
- l and u are terms (i.e. variables or constants) representing non-negative integers. If not specified, the defaults are 0 respectively ∞.
- $\prec_l$ and $\prec_u$ are comparison operators.

An aggregate literal is either an aggregate atom φ or its negation *not* φ. An *aggregate atom* is of the form

$$l \prec_l \#op\{t_{1_1}, \ldots, t_{1_m} : l_{1_1}, \ldots, l_{1_n}; \ldots; t_{o_1}, \ldots, t_{o_p} : l_{o_1}, \ldots, l_{o_q}\} \prec_u u$$

Most syntactical parts of aggregate literals are the same as for cardinality atoms, except that

- a head of a conditional literal is a tuple of terms $t_{i_1}, \ldots, t_{i_j}$ and
- $\#op$ is an aggregate function in $\{\#min, \#max, \#count, \#sum\}$.

Generally, a *rule* is of the form

$$h \leftarrow b_1, \ldots, b_m, not\ b_{m+1}, \ldots, not\ b_n.$$

where

- $h, b_1, \ldots, b_m$ are atoms (i.e. positive literals),
- $not\ b_{m+1}, \ldots, not\ b_n$ are negative literals,
- $H(r) = \{h\}$ is called the *head* of the rule,
- $B(r) = \{b_1, \ldots, b_m, \ldots, not\ b_{m+1}, \ldots, not\ b_n\}$ is called the body of the rule,
- $B^+(r) = \{b_1, \ldots, b_m\}$ is called the positive body of the rule and
- $B^-(r) = \{not\ b_{m+1}, \ldots, not\ b_n\}$ is called the negative body of the rule.

A rule r with $H(r)$ including a cardinality atom is called *choice rule*. A rule r where $B(r) = \{\}$, e.g. '$a \leftarrow$' is called *fact*. For facts, typically '$\leftarrow$' is omitted. A rule r where $H(r) = \{\}$, e.g. '$\leftarrow b$', is called *integrity constraint*, or simply *constraint*.

We allow the typically built-in arithmetic functions $(+, -, *, /)$ and comparison predicates $(=, \neq, <, >, \leq, \geq)$. For example, $A = B + C$ could also be rewritten as $=(A, +(B, C))$.

2.2 Semantics of ASP

The semantics of a non-ground ASP program is defined with respect to (w.r.t.) its *grounding*. A program's grounding can be defined in terms of its Herbrand universe and base. The *Herbrand universe* HU_P of a program P is the set of all constants appearing in P.

The grounding for a rule r without cardinality atoms and aggregates is the set of rules obtained by applying all possible substitutions of variables in r with constants in HU_P. The grounding of a rule which contains cardinality or aggregate literals is defined by the two-step instantiation described in [27]: first produce a set of partially grounded rules by substitution of variables occurring outside the cardinality/aggregate literal and then, within each partially grounded rule, replace each conditional literal by *a set of ground conditional literals* in which the remaining variables inside the cardinality or aggregate literal are substituted with constants in HU_P.

The *grounding* P_G of a program P is the union of all rule groundings. The *Herbrand base* HB_P w.r.t P is the set of all positive NAF literals (i.e. classical atoms) that occur in P_G.

An interpretation I satisfies a (ground) positive NAF literal λ (written as $I \models \lambda$) iff $\lambda \in I$. A positive cardinality literal is satisfied by I iff the number of satisfied head literals in the cardinality atom satisfies the lower and upper bounds l and u w.r.t. the order relations $\prec_l$ and $\prec_u$. Both, bounds and comparison symbols are optional. By default, $0 \leq$ is used for the lower and $\leq \infty$ for the upper bound. A positive aggregate literal is satisfied iff the value returned by the aggregate function *#op* applied on the set of term tuples that satisfy their conditions does not violate the lower and upper bounds. Here, *#count* counts the number of distinct satisfied term tuples, and *#min*, *#max* and *#sum* are calculating the minimum, maximum or sum of the first terms in the distinct satisfied term tuples. A negative literal *not* ω is satisfied by I (written as $I \models$ *not* ω) iff ω is not satisfied by I.

A ground rule r is satisfied by I (written as $I \models r$) iff the head is satisfied or the body is not. The body of a rule is satisfied by I iff all literals in the body are satisfied. An empty body is always satisfied. The head of a rule is satisfied iff the literal in it is satisfied. An empty head is never satisfied. In particular, integrity constraints are satisfied iff the body is not satisfied, i.e. the constraint is not violated. A program P is satisfied by an interpretation I iff all rules in its grounding P_G are satisfied.

An answer set for a program can be defined on the basis of the program's *reduct* [13, 27]. The reduct P^I of a ground program P relative to an interpretation $I \subseteq HB_P$ is defined as $P^I := \{H(r) \leftarrow B(r)^+ : r \in P, I \models B(r)^-\}$.

An interpretation $I \subseteq HB_P$ (which may be empty) is an *answer set* for a program P not containing choice rules iff

- I satisfies all rules r in P^I, i.e. $\forall r \in P^I : I \models r$ and
- I is subset-minimal, i.e. there is no $I' \subset I$ so that I' satisfies all rules in $P^{I'}$.

Choice rules can produce answer sets that are not subset-minimal, which leads to a slight change of semantics when such rules are present. For example, the program consisting only of the choice rule $\{a\}$. possesses the two answer sets $\{\}$ and $\{a\}$. In order to be in line with the original semantics and thus restore subset-minimality an equivalent program can be produced by extending the program as follows:

For every head a_i within a cardinality atom of a choice rule a new atom a_i' is introduced, which is not occurring elsewhere in the program. Furthermore, additional rules are added which assure that either a_i or a_i' but not both must be in an answer set. Thus, informally speaking, a_i' expresses that a_i is not in the interpretation. This way, the choice rule $\{a\}$. equivalently produces the two answer sets $\{a'\}$ and $\{a\}$, i.e. in the one answer set a_i is existent in the other it is not. For details, consult [11].

An ASP program is *unsatisfiable* iff it has no answer sets and *satisfiable* otherwise.

2.3 Constraint Satisfaction and Constraint ASP

A constraint satisfaction problems (CSP) can be defined as a three-tuple $\langle V, D = \{dom(v) : v \in V\}, C\rangle$ whereby V is a set of variables,[4] D is the set of domains of the variables in V and C is a set of constraints on variables in V. A solution to a CSP is an assignment $\forall v \in V : v := d, d \in dom(v)$ such that all constraints $c \in C$ are fulfilled. A CSP comprising only finite domains is called finite. If all domains are defined over discrete values (most commonly integers), the CSP is called discrete.

For integrating CP into ASP there are basically two approaches. First, solvers like Clingcon [21] are based on the extension of the ASP input language in order to support the definitions of constraints. Take as a simple example the following encoding in Clingcon (':-' represents left-implication and '!=' represents $\neq$):

```
num(1).num(2).num(3).
$domain(1..6).
var(X)$+var(Y)$+var(Z)$==6:-
                    num(X),num(Y),num(Z),X!=Y,Y!=Z,X!=Z.
var(1) $> 1.
$distinct{var(N):num(N)}.
```

The above encoding expresses that the sum of the three CSP variables *var*(1), *var*(2) and *var*(3) must be equal to six. The domain of the variables is 1..6. Furthermore, *var*(1) must be greater than one and all CSP variables must be distinct to each other. The ASP and CSP are fully integrated into one language. CSP specific constructs are indicated by $, like $+ or $==. *$distinct* constitutes a well known global constraint, i.e. a constraint over a set of variables. In Clingcon CSP variables are not defined explicitly but indirectly by the constraints. CP solving is integrated in the answer set production process and carried out by the Gecode solver. For more information on Clingcon please refer to [21].

The Ezcsp solver [2] is based on a different approach where ASP and CP are not integrated into one language. ASP rather acts as a specification language for Constraint Satisfaction Problems (CSPs). The main idea is that answer sets constitute

[4]Constraint variables are not to be confused with first-order logic variables in ASP. In the rest of this paper the word 'variable' refers to a constraint variable.

CSP encodings which are used as input for a CP solver. The above example can be expressed in Ezcsp as:

```
num(1).num(2).num(3).
cspdomain(fd).
cspvar(var(N),1,6):-num(N).
required(var(X)+var(Y)+var(Z)==6):-
                    num(X),num(Y),num(Z),X!=Y,Y!=Z,X!=Z.
required(var(1) > 1).
required(all_distinct([var/1])).
```

After some pre-processing, an answer set is calculated that includes *cspvar*, *required* and *cspdomain* facts. *cspdomain(fd)* denotes that the CSP is finite and discrete. Ezcsp is also able to handle real domains. CSP variables are explicitly defined by *cspvar* facts also specifying lower and upper bounds of the variable domains. Constraints are represented as *required* facts. For expressing global constraints, which requires referring to sets of CSP variables, Ezcsp allows the usage of functional symbols. E.g. $[var/1]$ designates all variables formed by the unary function *var*. Once an answer set has been produced, the CSP encoded by the *cspdomain*, *cspvar* and *required* facts is passed to the CP solver. As answer set production and CSP solution search are two separated processes, different CP solvers can be used in Ezcsp. Currently, Sicstus- and B-Prolog are supported.

The semantics of a program builds on the notion of *extended answer sets* [2]: A pair $\langle A, S\rangle$ is an extended answer set of program Π iff A is an answer set of Π and S is a solution to the CSP defined by A. We further define that the empty CSP (i.e. without any CSP variables) possesses the empty solution.

For CSP solution search, Ezcsp provides different search strategies impacting the underlying CP solver. In case of Sicstus Prolog as a CP solver, the built-in value selection strategies *step* (min domain value, when ascending order is used, max domain value when descending order is used) and *bisect* (bisection of the domain in the middle) are available. Similarly in case of B-Prolog, the bisection strategies *split* and *reverse_split* are supported. The supported variable selection strategies are *leftmost* (leftmost variable), *min* (leftmost variable with minimal lower bound), *max* (leftmost variable with maximal upper bound), and *ff* (first-fail). By the special *label_order/2* predicate it is also possible to define problem-dependent CSP variable orderings for the CP solver. However, there are no constructs for expressing problem dependent value or pruning strategies.

3 ASCASS—A Simple Constraint Answer Set Solver

In the following we introduce our novel CASP solver ASCASS. First, we give a brief introduction to the overall architecture of ASCASS and the encoding of CSPs in ASCASS. After that, we present the means for formulating problem-dependent heuristics.

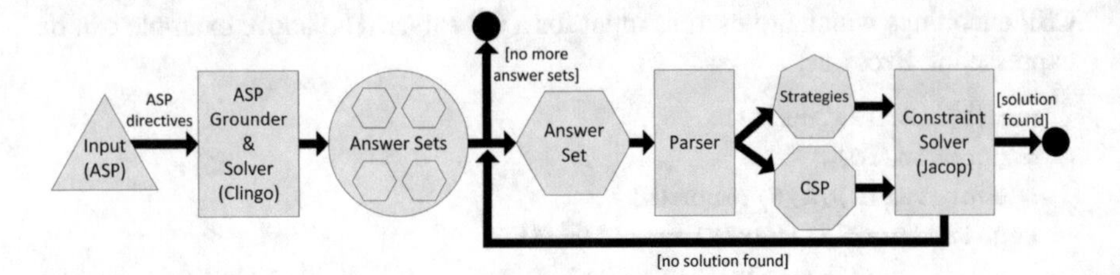

Fig. 1 Architecture of ASCASS

3.1 *Architecture*

ASCASS[5] is a finite discrete CASP solver following the approach of Ezcsp, i.e. the input language is pure ASP and the answer sets encode CSPs. Figure 1 shows the overall architecture of ASCASS. Answer set production (grounding and solving) is done by Clingo,[6] which is currently one of the most powerful ASP systems. The input language is the ASP standard ASP-Core-2.[7]

After answer set solving, a generated answer set is handed over to a parsing module that extracts the facts which encode the CSP and search directives. This information is used to instantiate a corresponding CSP in the CP solver and perform search conforming to the given search directives. Currently, Jacop[8] is used within ASCASS as a CP solver. In case that the CSP could not be solved by the CP solver or a timeout occurred (defined by the special predicate *csptimeout*(Δ)), the process continues with the next answer set, until a solution is found, or there are no more answer sets. The empty CSP (i.e. when there is not a single CSP variable) is always satisfiable and possesses the empty CSP solution.

3.2 *Encoding of CSPs*

ASCASS focuses on finite discrete Constraint satisfaction problems (CSPs). In order to encode a CSP within ASCASS there can be used a number of specific predicates. The following explanations refer to their grounded form.

The predicates *cspvar*($\alpha, \lambda, \upsilon$) and *cspvar*($\alpha, \lambda, \upsilon, \eta$) are responsible for encoding CSP variables. Hereby, α represents the variable name and λ and υ represent respectively the numerical lower and upper bound of the variable's domain. For example *cspvar*(x, 1, 10) stands for a CSP variable v with the domain [1..10]. The numerical priority η is used to define a custom variable selection ordering. When using the

[5]http://isbi.aau.at/hint/ascass.

[6]sourceforge.net/projects/potassco/files/clingo.

[7]http://www.mat.unical.it/aspcomp2013/files/ASP-CORE-2.03b.pdf.

[8]http://jacop.osolpro.com.

variable selection strategy *priority* (see below), the CP solver selects the variable with the highest priority first.

The predicate $cspconstr(\alpha, \rho, \tau)$ encodes a relational constraint (i.e. $=, <>, <, <=, >, >=$) over a variable α. ρ denotes the type of relation and must be a constant out of $\{eq, neq, lt, lteq, gt, gteq\}$. τ denotes another CSP variable or a numerical constant. For example, $cspconstr(x, lt, 5)$ expresses that variable x must be lower than 5.

The predicate $csparith(\alpha, \pi, \beta, \rho, \gamma)$ encodes arithmetic constraints. α, β and γ are CSP variable names. Like for *cspconstr*, the constant ρ denotes the type of relation. π is a constant representing an arithmetic operation. Currently, ASCASS supports addition (*plus*), subtraction (*minus*), multiplication (*mult*), division (*div*) and exponent (*exp*). For example, $csparith(xa, plus, xb, eq, xc)$ states that the sum of the values of *xa* and *xb* must be equal the value of *xc*.

For expressing logical constraints predicates of the form $cspif(\Xi_1, and, \Xi_2, and, \dots, and, \Xi_m, then, \Xi_{m+1}, or, \Xi_{m+2}, or, \dots, or, \Xi_n)$ can be used. Each Ξ consists of a variable α, a relational symbol ρ and another variable or numerical constant τ. For example, $cspif(x, lt, 5, and, y, gt, 10, then, z, gteq, 0)$ is to be read as 'if x is lower than 5 and y is greater than 10 then z must be non-negative'.

Global constraints are constraints over arrays of variables. In ASCASS global constraints are defined by predicates of the form $cspglobal(\sigma_1, \dots, \sigma_m, \kappa)$ and $cspglobal(\sigma_1, \dots, \sigma_m, \kappa, \tau_1, \dots, \tau_n)$. κ is a constant denoting the type of global constraint. $\sigma_1, \dots, \sigma_m$ represent arrays of variables. $\tau_1, \dots, \tau_n$ represent single CSP variables or integers. The selection of global constraints currently supported by ASCASS has been determined by the needs of our application areas and will be further expanded. ASCASS currently supports the following global constraints[9]:

- min: $cspglobal(\sigma, min, \tau)$, the minimum value of the variables σ is equal to τ
- max: $cspglobal(\sigma, max, \tau)$, the maximum value of the variables σ is equal to τ
- sum: $cspglobal(\sigma, sum, \tau)$, the sum of values of the variables σ is equal to τ
- count: $cspglobal(\sigma, count, \tau_1, \tau_2)$, τ_1 is equal to the counted number of variables in σ with value τ_2
- global cardinality: $cspglobal(\sigma_1, \sigma_2, gcc)$, a more general counting constraint where the occurring values in σ_1 are counted in the corresponding counter variables in σ_2
- all different: $cspglobal(\sigma, alldiff)$, all variables in σ are mutually unequal
- element: $cspglobal(\sigma, element, \tau_1, \tau_2)$, the value of the τ_1-th variable in σ is equal to τ_2
- cumulative: $cspglobal(\sigma_1, \sigma_2, \sigma_3, cumulative, \tau)$, σ_1 represents the starting times of $|\sigma_1|$ many jobs, σ_2 represents the durations of the jobs, σ_3 represents the amounts of needed resources of the jobs and τ represents the allowed accumulated amount of resources at any time point
- bin packing: $cspglobal(\sigma_1, \sigma_2, \sigma_3, binpacking)$, σ_1 represents bin assignments for $|\sigma_1|$ many items, σ_2 represents the bin sizes of the $|\sigma_2|$ many bins and σ_3 represents the item sizes

[9]More information about global constraints can be found at http://jacop.osolpro.com/guideJaCoP.pdf and http://sofdem.github.io/gccat/.

v(1,1)	v(1,2)	...	v(1,c)	v(1,all)
v(2,1)	v(2,2)	...	v(2,c)	v(2,all)
...	...	...	...	...
v(r,1)	v(r,2)	...	v(r,c)	v(r,all)
v(all,1)	v(all,2)	...	v(all,c)	v(all,all)

Fig. 2 Concept of variable arrays in ASCASS

In order to address arrays of CSP variables, ASCASS not only allows simple constants but also n-ary functional terms for variable names of the form $\phi(\iota_1, \dots, \iota_n)$ with $\iota_1, \dots, \iota_n$ representing string or integer arguments (see Fig. 2). The special functional argument *all* acts as a placeholder and can be used for addressing arrays of variables. For example, take the four variable definitions $cspvar(v(1,1), 1, 10)$, $cspvar(v(1,2), 1, 10)$, $cspvar(v(2,1), 1, 10)$ and $cspvar(v(2,2), 1, 10)$. A natural interpretation of the arguments is *row* and *column* of a two-dimensional variable array. Consequently, $cspglobal(v(all, 2), alldiff)$ expresses that the values of all second column's variables, in our case $v(1,2)$ and $v(2,2)$, must be different to each other. *v(all, all)* stands for all variables in the two-dimensional array, i.e. all variables formed by the functional symbol v with arity 2.

3.3 Encoding of Variable Selection Strategies

Apart from the predicates for defining a CSP, ASCASS provides predicates for steering the search of the CP solver. The predicates *cspvarsel*(ε) and *cspvarsel* (ε, θ) define the variable selection strategy to be used. Herby, ε is the primary selection strategy and, if defined, θ acts as a secondary, tiebreaking strategy. For variable selection, ASCASS currently supports the problem-independent built-in strategies *smallestDomain*, *mostConstrainedStatic*, *mostConstrainedDynamic*, *smallestMin*, *largestDomain*, *largestMin*, *smallestMax*, *maxRegret*, *weightedDegree* and the problem-dependent strategy *priority*. The default for ε is *smallestDomain* and the default for θ is *mostConstrainedDynamic*.

When using the *priority*-strategy, ASCASS builds an ordering of the CSP variables based on the provided priorities η in $cspvar(\alpha, \lambda, \upsilon, \eta)$. Variables with high priorities are selected first. Variables for which there is no η defined are selected as the last ones.

3.4 Encoding of Value Selection Strategies

For value selection ASCASS provides the predicates *cspvalsel*(ϕ) and *cspvalsel* (ϕ, φ) where ϕ and φ are constants denoting the strategy. As it is often important to have different value selection strategies for different sets of variables, ASCASS provides also the predicates *cspvalsel*(σ, ϕ) and *cspvalsel*(σ, ϕ, φ) where σ represents an array of variables like in global constraints. ASCASS supports the already built-in strategies *indomainMin*, *indomainMiddle*, *indomainMax* and *indomainRandom*. For expressing problem-dependent value selection strategies, the novel strategy *indomainPreferred* can be used.

When using *indomainPreferred*, the CP solver first tries to use specified values before changing to the built-in strategy φ (*minDomain* if not stated otherwise). For specifying preferred values, ASCASS provides the special predicate $cspprefer(\alpha, \rho, \tau)$ and $cspprefer(\alpha, \rho, \tau, \eta)$. Like for relational constraints, α represents a CSP variable, ρ represents a relational symbol and τ stands for a further variable or a numerical constant. For example, $cspprefer(v, eq, 5)$ states that for the CSP variable v a preferred value is 5. In order to specify an ordering of the specified values, it is possible to make use of a numerical priority η. Higher priority statements are taken into account first by ASCASS. For example, if there is given $cspprefer(v, eq, 5, 1)$ and $cspprefer(v, eq, 20, 2)$, ASCASS tries to first label v with 20 and only after that with 5. Of course, only preferred values are taken into account which are still in the variable's domain. In case that τ denotes another variable, the minimum value in the current domain of τ is used as a preferred value, i.e. τ does not need to be singleton for specifying a preferred value of α. This in combination with global constraints is a highly dynamic and powerful mechanism.

As with the relational constant *eq* in combination with the priorities η every ordering of preferred values can be expressed, the usage of *lt*, *lteq*, *gt* and *gteq* can be clearly seen as syntactic sugar. By using *lt*, *lteq*, *gt* and *gteq* sets of preferred values can be expressed:

- $lteq\ \tau : \{\tau, \tau - 1, \ldots, -\infty\}$
- $lt\ \tau : \{\tau - 1, \ldots, -\infty\}$
- $gteq\ \tau : \{\tau, \tau + 1, \ldots, \infty\}$
- $gt\ \tau : \{\tau + 1, \ldots, \infty\}$

Note that all preferred values of such a set P have the same priority (possibly given explicitly by η). For defining an order relation over P, i.e. fix the order in which ASCASS considers the preferred values in P, the following holds: For *lt* and *lteq* decreasing order is used, i.e. $\tau, \tau - 1, \ldots, -\infty$ and for *gt* and *gteq* increasing order is used, i.e. $\tau, \tau + 1, \ldots, \infty$. For example having the variable definition $cspvar(v, 1, 10)$ and the value selection strategy *cspvalsel*(*indomainPreferred*, *indomainMin*), $cspprefer(v, lt, 5)$ would effect that ASCASS considers the domain values in the following order: 4, 3, 2, 1, 5, 6, 7, 8, 9, 10. The reason why for *lt* and *lteq* descending order and for *gt* or *gteq* ascending order is used is simply the following: Would it be the other way round, the behavior with *lt* and *lteq* would conform to *indomainMin* and with *gt* and *gteq* to *indomainMax*.

3.5 Encoding of Pruning Strategies

The third component of many problem-dependent heuristics is the pruning strategy. For specifying how a search tree is pruned, ASCASS provides the special predicate *cspsearch*(ω, μ). Hereby, ω specifies the pruning type and μ specifies a numerical limit that, when reached, triggers backtracking. Again it could be beneficial having different limits for different groups of variables or even having no limit on certain variables whilst search on others is limited. To this, ASCASS provides the predicate *cspsearch*(σ, ω, μ) with σ denoting an array of variables like for global constraints.

Currently, ASCASS provides two pruning types. *cspsearch*(*limited*, μ) limits the number of wrong decisions for variables. If the number μ of wrong choices for a variable is reached, backtracking is triggered and the counter for the variable is reset. For example, *cspsearch*(*limited*, 3) specifies that for every variable v there must not be more than three labeling trials for v within a search branch. The second pruning type is based on limited discrepancy search [14] and operates on the level of search paths. When specifying *cspsearch*(*lds*, μ) only a certain number of wrong decisions (called discrepancies) along the whole search path is allowed. If this number reaches μ, backtracking is triggered.

Furthermore, it is possible to limit search time of the CP solver by *csptimeout* (Δ) where Δ is the number of seconds when the timeout is triggered. The timeout concerns only the search of Jacop so that search might start over based on the next answer set if such exists.

3.6 Directives for Answer Set Production

In order to specify which heuristic is to be utilized by Clingo, the special ASCASS predicate *aspheuristic*(v) can be exploited. Hereby, v is a constant denoting the heuristic which is passed to Clingo as a command line option $--heur = v$. As this happens before the actual answer set solving, *aspheuristic*(v) must only be used as a single fact within program source code. Common heuristics are *VSIDS* or *Berkmin* [17]. When using *aspheuristic*(*domain*), Clingo harnesses a user-defined heuristic defined via the *_heuristic* predicate built-in Clingo [12]. For limiting the number of produced answer sets, the special predicate *aspnumas*(Δ) can be utilized. Δ is a non-negative integer and is passed to Clingo as a command line option. The default is ‘1’ and ‘0’ effects the production of all answer sets. Like *aspheuristic*, also *aspnumas* must only be used as a single fact within the problem source code. Similarly, *asptimeout*(Δ) specifies a timeout for answer set solving.

4 Application to System Configuration

The configuration of products and services [8, 26], or more generally the configuration of systems, is an important task in many production strategies such as mass customization, configure-to-order, or assembly-to-order. On the one hand, the basic goal is to provide customers with products and services which fulfill all their requirements and maximize the satisfaction of their preferences. On the other hand, these products and services shall be offered at mass production efficiency. In order to fulfill these goals, systems are assembled by pre-designed and pre-fabricated components where such components themselves may be assembled by components. In addition, there are means to individualize components by characterizing their properties by parameters (e.g. color or size).

The goal of the configuration task is to generate a system description (i.e. a configuration) for given requirements and preferences that (1) contains all the information needed for manufacturing or service provision in an explicit, succinct, and simple to process format (e.g. a set of facts) and (2) describes a system which will fulfill the requirements and will optimize the preferences. One of the most prominent approaches for solving the configuration task is knowledge-based configuration which employs a knowledge-base to describe all configurations for all requirements and preferences. Such a knowledge-base must be specific enough such that only working systems which fulfill the requirements will be provided. Moreover, the configuration knowledge-base must be as general as possible such that no opportunity for satisfying the requirements and optimizing the preferences is missed. Typically a configuration knowledge-base comprises a set of component types and descriptions representing various technical and non-technical constraints, e.g. which components must be selected and connected as well as how the parameters must be set such that the customer satisfaction is maximized.

4.1 The Partner Units Configuration Problem

The partner units problem (PUP) [7] is a perfect representative of a configuration problem in the classical sense, i.e. where certain components have to be connected so that predefined user requirements and technical constraints are respected [20]. Because of its generic nature it posses many real world application domains like railway safety, surveillance or electrical engineering [1, 28]. The PUP is $\mathscr{NP}$-complete in the general case and also for most industrially important subclasses. Furthermore, it is one of the hardest benchmark problems participating in the ASP competitions[10][34].

The PUP originates in the domain of railway safety systems. One of the problems in this domain is to make sure that certain rail tracks are not occupied by a train/wagon before another train enters this track. The signals for the corresponding occupancy indicators are calculated by special processing units based on the input of several observing sensors. Because of fail-safety and realtime requirements the number of sensors respectively indicators which can be connected to the same unit is limited (called unit capacity, UCAP). Also one sensor/indicator device can only be directly connected to one unit. However, a unit can be connected to a limited number (called inter unit capacity, IUCAP) of other units. These units are called the partner

[10]Further information can be found at http://www.mat.unical.it/aspcomp2014/.

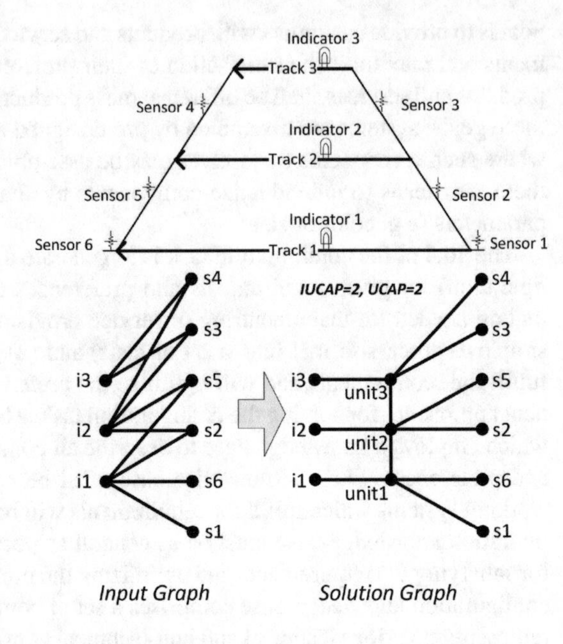

Fig. 3 Railway track layout, PUP input and solution

units of the unit. Devices (i.e. sensors and indicators) can only communicate with devices connected to the same unit and with devices connected to one of the partner units. Given the IUCAP, UCAP and a bipartite input graph represented by edges specifying which sensor data is needed in order to calculate the correct signal of an occupancy indicator, the problem consists in connecting sensors/indicators with units and units with other units such that all communication requirements are fulfilled and IUCAP and UCAP are not violated. Formally, the Partner Units Decision Problem (PUDP) can be defined as follows:

Given is a bipartite Graph $G = (I, S, E)$ and two natural numbers *IUCAP* and *UCAP*. The PUDP is to decide whether there is a partition of the vertices $I \cup S$ into a set U of units such that each unit contains at most *UCAP* vertices from I and at most *UCAP* vertices from S, and has at most *IUCAP* connected units. Two different units $U_1 \in U$ and $U_2 \in U$ are connected whenever $v_1 \in U_1$ and $v_2 \in U_2$ and $\{v_1, v_2\} \in E$.

For minimizing hardware costs, a common further objective is the minimization of the number of units, i.e. $|U| \rightarrow min$.

Figure 3 shows a simple example for a railway track layout, the corresponding bipartite input graph and a possible solution for IUCAP = 2 and UCAP = 2. In order to calculate the correct signal for Indicator 3 only data from Sensor 3 and Sensor 4 is needed. If the number of outgoing wheels counted by Sensor 4 is equal to the incoming wheel counts of Sensor 3 then Track 3 is empty. In order to calculate the correct signal for Indicator 2 it is not sufficient to only incorporate data from Sensor 2 and Sensor 5 as it is not clear whether a wheel has headed to or is coming from Track 3. Therefore, additional data from Sensor 3 and Sensor 4 is needed.

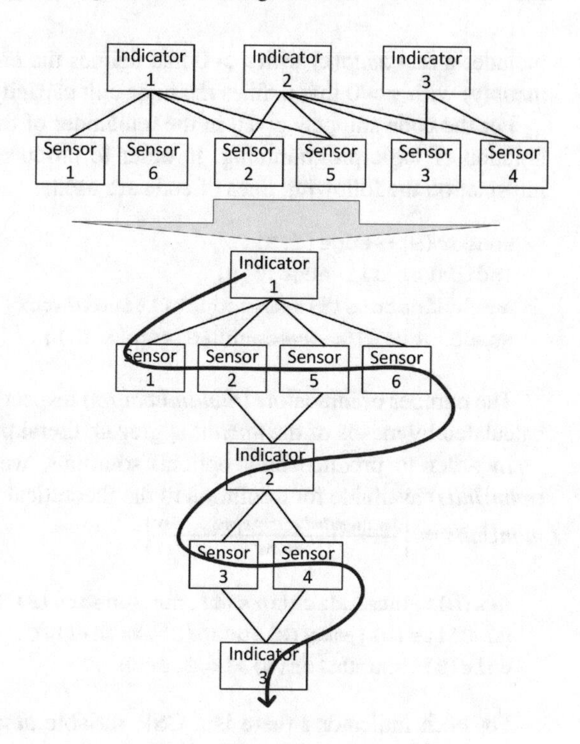

Fig. 4 Breadth-first restructured input graph for example in Fig. 3, (start node = *Indicator 1*)

The state-of-the-art heuristic for solving PUP is the QuickPup heuristic proposed in [33]. QuickPup is based on three major techniques. First, based on the input graph and a distinguished root indicator, QuickPup produces a topological ordering of the devices, which is basically the minimum distances from the root indicator to all other devices. The distance to itself is zero, the distance to the direct neighbors is one, the distance to the neighbors of the neighbors is two and so forth. This reflects the (partial) ordering in which the devices should be processed. Figure 4 shows the idea of the breadth-first order for the example in Fig. 3. Thus, when beginning with Indicator 1 as starting node the ordering would be: $Indicator1 \rightarrow Sensor1 \rightarrow Sensor2 \rightarrow Sensor5 \rightarrow Sensor6 \rightarrow Indicator2 \rightarrow Sensor3 \rightarrow Sensor4 \rightarrow Indicator3$.

Second, for each device, first try to place it on the next empty unit and if this is unsuccessful try the already used units in descending order. Third, try different root indicators, and consequently different topological orderings, and limit search for each trial. The intuition behind that is that not all root indicators are equally good to start search from.

4.2 Encoding in ASCASS

The input for a PUP encoding comprises a set of $egde(i, s)$ facts where i takes the numerical id of an indicator and s takes the id of a sensor. Additionally the input includes a fact *ucap*(x) with $x > 0$ that defines the unit capacity (UCAP) and a fact *iucap*(y) with $y > 0$ that defines the inter-unit capacity (IUCAP).

For the code snippets given in the remainder of this section we use the standard notation of logic programming. In order to produce explicit indicator and sensor information the following lines of code are used:

```
sensor(S):-edge(I,S).
indicator(I):-edge(I,S).
numIndicators(N):-N=#count{I:indicator(I)}.
numSensors(N):-N=#count{S:sensor(S)}.
```

The number of indicators (*numIndicators*) respectively sensors (*numSensors*) are calculated by means of the *#count* aggregate literal provided by Clingo.

In order to produce only optimal solutions, we restrict the number of units (*numUnits*) available for a solution to the theoretical lower bound, i.e.
$numUnits = \left\lceil \frac{max(numIndicators, numSensors)}{UCAP} \right\rceil$:

```
max(M):-numIndicators(E),numSensors(F),M=#max(E;F).
numUnits(N):-max(M),ucap(C),N=(M+1)/C.
unit(Z):-numUnits(N),1<=Z,Z<=N.
```

For each indicator i there is a CSP variable *device*(i, 1) and for each sensor s there is a CSP variable *device*(s, 2). This way it is also possible to refer to the array of all CSP device variables as *device*(*all*, *all*), to only the indicator variables as *device*(*all*, 1) and to the sensor variables as *device*(*all*, 2) which will be useful later. The value range for these CSP variables is [1..*numUnits*]. Furthermore, the variables get a priority defining the topological order in which they are labeled by ASCASS:

```
cspvar(device(I,1),1,N,P):-numUnits(N),iPriority(I,P).
cspvar(device(S,2),1,N,P):-numUnits(N),sPriority(S,P).
```

The calculation of the priorities is explained in detail below.

In order to assure UCAP, for each unit u there are two counting variables *ci*(u) and *cs*(u). These variables can take values in the range [0..*UCAP*]. Furthermore, for each unit u there are two *count* global constraints counting the number of indicator respectively sensor variables taking the value u:

```
cspvar(ci(U),0,C):-ucap(C),unit(U).
cspvar(cs(U),0,C):-ucap(C),unit(U).
cspglobal(device(all,1),count,ci(U),U):-unit(U).
cspglobal(device(all,2),count,cs(U),U):-unit(U).
```

In order to capture which unit $u1$ is connected to which unit $u2$ there are $numUnits \times numUnits$ many CSP variables (i.e. *conn*($U1$, $U2$)). The variables can take values in the range [0..1] if $u1 <> u2$. Otherwise, the variables' ranges consists of only a single value, i.e. [1..1]. This is because in our model each unit u is always connected to itself. Furthermore, there is a constraint assuring symmetry, i.e. if $u1$ is connected to $u2$ also $u2$ is connected to $u1$:

```
cspvar(conn(U1,U2),0,1):-unit(U1),unit(U2),U1<>U2.
cspvar(conn(U,U),1,1):-unit(U).
cspconstr(conn(U1,U2),eq,conn(U2,U1)):-
                                unit(U1),unit(U2),U1<U2.
```

For summing up how many units are connected to a unit u we make use of the global *sum* constraint. The used summing variables can hereby take values in the range $[1..IUCAP + 1]$ as every unit is also connected to itself:

```
cspvar(sumconns(U),1,K+1):-iucap(K),unit(U).
cspglobal(conn(U,all),sum,sumconns(U)):-unit(U).
```

In order to make the summing variables and constraints take effect, it must be assured that any connection variable $conn(u1, u2)$ is set to one whenever there is an $edge(i, s)$ in the input so that $device(i, 1) = u1$ and $device(s, 2) = u2$. Following the approach of [6], this is implemented by means of the global *element* constraint. Given an array of CSP variables *arr*, an index i and a value v, an *element* constraint assures that the ith variable in *arr* is equal to v. In our case, for each $edge(i, s)$ in the input there is such a global constraint setting the appropriate connection variable within $conn(all, all)$ to one:

```
cspglobal(conn(all,all),element,index(I,S),1):-
                                             edge(I,S).
```

As the *element* constraint cannot directly handle multi-dimensional arrays, the respective index is calculated as:

```
index(i,s)=(device(i,1)-1) x numUnits + device(s,2)
```

The formulation with constraints is straightforward.

The priorities for the device variables (i.e. $device(i, 1)$ and $device(s, 2)$ are based on a topological ordering of the devices. Given the layer of a sensor or indicator whereby the root of the topological graph is at layer zero, the priority is higher the lower the layer is:

```
iPriority(I,P):-indicatorLayer(I,L),P=-L.
sPriority(S,P):-sensorLayer(S,L),P=-L.
```

The effect is that, given a root indicator, ASCASS first tries to label the root indicator, then the neighbors of the root indicator, then the neighbors of the neighbors, and so on. In our implementation a choice rule is used to express that there is exactly one distinguished indicator that acts as root. This indicator is always placed at the first unit:

```
1{root(I):indicator(I)}1.
cspconstr(device(I,1),eq,1):-root(I).
```

The choice rule $1\{root(I) : indicator(I)\}1$ produces one answer set for each root indicator and asserts a *root*(*i*) fact.

For calculating the actual layers, we first calculate the minimum distances to the root whereas root indicator has a zero distance to itself. The maximum possible distance is equal to the total number of devices:

```
indicatorDist(I0,0):-root(I0).
sensorDist(S,D+1):-indicatorDist(I,D),edge(I,S),
                                    numDevices(M),D<M.
indicatorDist(I,D+1):-sensorDist(S,D),edge(I,S),
                                    numDevices(M),D<M.
numDevices(M):-numIndicators(E),numSensors(F),M=E+F.
```

The layers are calculated by using the *#min* aggregate literal from Clingo:

```
indicatorLayer(I,Dmin):-indicator(I),
                  Dmin=#min{D:indicatorDist(I,D)}.
sensorLayer(S,Dmin):- sensor(S),
                   Dmin = #min{D:sensorDist(S,D)}.
```

First to try to place devices on unused units and, only if not successful, on used units in descending order can be expressed in ASCASS by means of preferred values:

```
cspprefer(device(I,1),lteq,nextUnit):-indicator(I).
cspprefer(device(S,2),lteq,nextUnit):-sensor(S).
```

The CSP variable *nextUnit* points to the next unused unit, which is the current unit plus one[11]:

```
cspvar(curUnit,1,N):-numUnits(N).
cspvar(nextUnit,1,N+1):-numUnits(N).
csparith(curUnit,plus,one,eq,nextUnit).
```

For the calculation of the current unit, i.e. the highest number taken by some *device*$(i, 1)$ or *device*$(s, 2)$ variable, the global *max* constraint is used:

```
cspglobal(device(all,all),max,curUnit).
```

[11] Within the constraint, the helping variable *cspvar*$(one, 1, 1)$ is used as arithmetic constraints only accept variables in ASCASS.

As ASCASS uses the lower bound of variables for calculating the preferred values, each device variable is first tried to be bound to values lower than or equal to the lower bound of $nextUnit = curUnit + 1$ in descending order.

In order to control how many units are maximally tried per device variable, the search is pruned such that only the next unit and a limited number of already used units can be tried before backtracking is triggered. In our implementation we use the following statement for only trying the next, the current and the last unit:

```
cspsearch(limited,3).
```

We furthermore restrict the maximum CSP search time for each call of the CP solver in order to try different start indicators:

```
csptimeout(300).
```

In order to enforce that ASCASS respects the problem-dependent strategies, *cspvarsel*(*priority*) and *cspvalsel*(*device*(*all*, *all*), *indomainPreferred*) must be included. Thus, the concepts of QuickPup can be fully expressed in a declarative way by ASCASS. To the best of our knowledge, this is not possible within any other ASP or CASP approach.

4.3 Evaluation

We tested the ASP solver Clingo 4 and the CASP solvers ASCASS, Clingcon and Ezcsp on the PUP benchmark suite used in [1].[12] Clingo was tested using the PUP encoding proposed in [1].[13] Only optimal solutions were allowed. The tests were run on a 3.2 GHz machine with 64 GB of RAM, assuring that the grounding bottleneck does not play a role for the tested instances[14] and performance can be attributed to the search phase.

In the Clingcon model, CSP variable selection, value selection or pruning strategies cannot be manipulated. For Ezcsp, it is possible to express the topological variable orderings similar to ASCASS. However, there are no means for pruning search or problem-dependent value strategies.

Table 1 depicts how many instances of each type in the benchmark suite could be solved by the different approaches within a 1000 s time frame. Clingo using VSIDS heuristic peformed very well on the benchmark suite showing once again that the conflict-driven search techniques employed by Clingo are quite powerful. Also Ezcsp was able to solve some instances. Using other built-in heuristics did not result in

[12]Encodings and benchmark instances can be found at http://isbi.aau.at/hint/ascass.

[13]The 'new' encoding provided by the ASP competition 2014 was found to be inconsistent as it also produces answer sets for unsatisfiable instances.

[14]The biggest grounding in the ASP model was ~ 12 GB.

Table 1 Solved instances within 1000 s

	#	Clingo	ASCASS	Clingcon	Ezcsp
Double (IUCAP = 2)	10	2	10	0	2
Doublev (IUCAP = 2)	6	3	6	0	0
Triple (IUCAP = 2)	3	2	3	0	2
Triple (IUCAP = 4)	7	6	6	0	3
Grid (IUCAP = 4)	10	10	10	0	0
Total	36	23	35	0	7

better performance. Clingcon was not able to solve a single instance. In the contrary, ASCASS was able to solve all but one instances within time limits. We want to point out that only optimal solutions (i.e. minimum number of units) were allowed for easing the grounding bottleneck of conventional ASP. Increasing the number of allowed units in a solution would increase grounding size for ASP significantly. In the cases of ASCASS and Ezcsp, increasing the number of allowed units would not affect the grounding size as the number of allowed units is captured by the upper bounds of the CSP variables.

We want to make clear that the superior performance of ASCASS can be attributed to the inclusion of the QuickPup strategies. This was crosschecked by removing the heuristic parts from the ASCASS problem encodings. It is to be noted that Quick-Pup originally was designed for producing only near-optimal solutions. However, the concepts of QuickPup obviously also work well for finding optimal solutions.

5 Application to Production Scheduling

The scheduling of jobs [5] is an important task in almost all production systems in order to optimize various objectives such as resource consumption, tardiness, or flow time. In general, jobs are structured into operations which must be allocated to resources such that various requirements are satisfied and in addition an objective function is optimized. Research on scheduling has developed an extensive classification scheme in order to characterize different scheduling problems. In this paper we focus on the job-shop scheduling problem.

5.1 The Job-Shop Scheduling Problem

The job-shop scheduling problem (JSP) is among the most famous $\mathcal{NP}$-hard [10] combinatorial problems and can be defined as follows:

- Given is a set $M = \{machine_1, \dots, machine_m\}$ of machines and a set $J = \{job_1, \dots, job_j\}$ of jobs.
- Each job $j \in J$ consists of a sequence of operations $O_j = \langle j_1, \dots, j_{l_j} \rangle$ whereby j_{l_j} is the last operation of job j.
  Practically, jobs can be interpreted as products and operations can be interpreted as their production steps. With respect to a job j and its operation j_i, the operation j_{i+1} is called successor and the operation j_{i-1} is called predecessor.
- Each operation o has an operation length $length_o \in \mathbb{N}$.
- Each operation o is assigned to a machine $machine_o \in M$ by which it is processed.
- A (consistent and complete) schedule consists of a starting time $start_o$ for each operation o such that:
  - An operation's successor starts after the operation has been finished, i.e. with respect to a job j and the operations j_i and j_{i+1}:
    $start_{j_{i+1}} \geq start_{j_i} + length_{j_i}$
  - Operations processed by the same machine are non-overlapping, i.e. with respect to two operations $o1 \neq o2$ with $machine_{o1} = machine_{o2}$:
    $start_{o1} \neq start_{o2}$
    $start_{o1} < start_{o2} \rightarrow start_{o1} + length_{o1} \leq start_{o2}$
- Makespan, i.e. the time period needed for processing all operations, is minimized. I.e.:
  - $\max_{j \in J, o \in O_j} \{start_o + length_o\} \rightarrow min$

Many approaches have been used to solve scheduling problems like tabu and large neighborhood search [35], simulated annealing or genetic algorithms [22]. Besides such meta heuristic also constraint based approaches have a long and successful history in solving scheduling problems [4].

The declarative representation as a constraint satisfaction problem [23] which is solved by some general purpose constraint solver has the big practical advantage that adaptions of the problem specification can be made quickly and easily compared to changing imperative code in productive use. This is of special importance in nowadays manufacturing environments offering highly dynamic fabrication regimes such as mass customization, just-in-time or lean production [25]. Moreover the encountered scheduling problems most often do not fully conform to idealized scheduling problems found in literature and as a matter of fact almost every real world scheduling problem has specific characteristics regarding the manufacturing processes. Furthermore, scheduling problems might diversify frequently due to change of the production infrastructure, availability of operating staff, order situation or product portfolio. Constraint-based approaches fit this need of flexibility in that special problem characteristics can be rigorously expressed by adding or removing constraints or variables.

However, in real world environments often neither of the above mentioned approaches is applied as problem instances are simply too large. Common problem instances for a weekly workload in semi-conductor domains like those of our project partner Infineon Austria Technologies are of the order of 10^4 operations on 10^2 machines in the back-end, i.e. where the products are made ready for shipping, and 10^5 operations on 10^3 machines in the front-end, i.e. where the chips are actually produced. Consequently, runtimes beyond quadratic complexities are not acceptable.

One widely employed state-of-the-art technique for dealing with such large scheduling problem instances in nowadays manufacturing environments is the application of dispatching rules [15]. Dispatching rules are greedy heuristics for step-wise deciding which is the operation to be scheduled next. There are basically two ways for using dispatching rules. First, dispatching rules can be directly applied by operator staff for steering the production process. As soon as a machine becomes idle, an operator decides which of the runnable operations in the dispatch list of the machine is to be loaded next. Second, schedules are built by doing simulations based on different dispatching rules. A schedule which is found to be good enough is then carried out. The big advantage of simulation-based scheduling compared to the direct application in the production process is that it is possible to predict when a product will be finished. This is especially important for customer relationship management.

In following, we will show how the JSP and some dispatching rules can be represented in a HCASP framework in a strongly declarative way. We conform hereby to the language of ASCASS. Translation into the language of similar frameworks like [2] is straight-forward.

5.2 Encoding in ASCASS

We expect the input for our JSP encoding to consist of logic facts of the following form:

- time(t): max time horizon is t
- job(j): there is a job j
- machine(m): there is a machine m
- jobOperation(j, op): operation op belongs to job j
- opLength(op, l): operation op has length l
- opMachine(op, m): operation op is to be processed by machine m
- precedes(op1, op2): operation op2 is the successor of operation op1

For each operation op that is to be processed by machine m we define a CSP variable for the starting time. The lower bound of the corresponding domain is 0 and the upper bound is set to T whereby T captures the value t specified by the time-fact in the input:

```
cspvar(start(Op,M),0,T):-opMachine(Op,M),time(T).
```

Likewise, there are CSP variables for the operation lengths and the finishing times:

```
cspvar(length(Op,M),L,L):-opMachine(Op,M),
                                   opLength(Op,L).
cspvar(finished(Op,M),0,T):-opMachine(Op,M),time(T).
```

Arithmetic constraints link together the starting times, lengths and finishing times:

```
csparith(start(Op,M),plus,length(Op,M),
eq,finished(Op,M)):-opMachine(Op,M).
```

For each precedence fact in the input there is a primitive constraint expressing that the finishing time of a preceding operation must be lower or equal than the start time of its successor:

```
cspconstr(finished(Op1,M1),lteq,
start(Op2,M2)):-opMachine(Op1,M1),opMachine(Op2,M2),
                              precedes(Op1,Op2).
```

In order to enforce non-overlapping of the operations that are processed by the same machine we make use of the global 'cumulative' constraint[15]:

```
cspglobal(start(all,M),length(all,M),
consumption(all,M),cumulative,resources(M)):-
                                        machine(M).
```

Hereby, resource variables capture the information that each machine can only process a single operation at a time and consumption variables capture the information that also all operations need exactly a single resource at a time:

```
cspvar(resources(M),1,1):-machine(M).
cspvar(consumption(Op,M),1,1):-opMachine(Op,M).
```

Finally, to restrict the makespan there is a corresponding CSP variable where its domain's upper bound is set to the value specified by the time fact in the input. The makespan variable is used whithin a global constraint calculating the maximum of the finishing times of all operations on all machines:

[15] http://www.web.emn.fr/x-info/sdemasse/gccat/Ccumulative.html. The special constant 'all' is used in the ASCASS system to refer to arrays of CSP variables.

```
cspvar(makespan,0,T):-time(T).
cspglobal(finished(all,all),max,makespan).
```

The extension of the *basic* encoding in order to realize dispatching rules is straight forward by defining CSP variable priorities for the start variables. In a HCASP framework such priorities can be used for specifying in which order the CSP variables are processed by the CP solver. In ASCASS the priorities are syntactically integrated in the cspvar-predicate[16]:

```
cspvar(start(Op,M),0,T,P):-opMachine(Op,M),time(T),
                                          priority(Op,P).
```

A simple dispatching rule might just process all first job operations at first, then process all second job operations and so on. Priorities for that can be expressed as:

```
seq(Op,1):-operation(Op),not precedes(_,Op).
seq(Op2,S2):-precedes(Op1,Op2),seq(Op1,S1),S2=S1+1.
priority(Op,P):-seq(Op,S),P=-S.
```

Hereby, the seq/2 predicate specifies the sequence number of the operations, i.e. at which step in the job the operation is carried out. We will refer to this rule as SEQ. The priorities are simply the inverted sequence numbers, i.e. the smaller sequence numbers result in higher priorities. Respecting the variable priorities, the labeling order of the underlying CP solver conforms to the dispatching rule.

One of the most effective dispatching rules for minimizing the makespan is the most-total-work-remaining (MTWR) rule [16]. According to MTWR, the next operation to be dispatched belongs to a job such that the sum of lengths of all remaining operations is maximal. The priorities according to MTWR can be expressed as:

```
priority(Op,L):-opLength(Op,L),not precedes(Op,_).
priority(Op1,P2+L1):-precedes(Op1,Op2),
                    opLength(Op1,L1),priority(Op2,P2).
```

The idea is that the last operations of jobs, i.e. those with no successor, have a priority equal to their operation length. For any other operation the priority is calculated as the sum of the operation length and the priority of the successor.

5.3 Evaluation

The purpose of this evaluation is to illustrate the efficacy of dispatching rule inspired CASP encodings. For evaluation purposes we produced test instances of realistic

[16]In Ezcsp there is a special label_order predicate to be used.

Table 2 Solved instances within 3600 s

	#	Basic	SEQ	MTWR
long-jobs (max makespan = 600k)	12	11	12	11
long-jobs (max makespan = 800k)	12	1	1	5
long-jobs (max makespan = 1000k)	12	4	8	11
long-jobs (max makespan = 1200k)	12	7	11	12
short-jobs (max makespan = 600k)	12	0	2	2
short-jobs (max makespan = 800k)	12	4	12	11
short-jobs (max makespan = 1000k)	12	3	12	12
short-jobs (max makespan = 1200k)	12	3	12	12

sizes (up to 10,000 operations of up to more than 2000 jobs on up to 100 machines) which are on the one hand patterned on problems of our project partner Infineon Austria Technologies and on the other hand have the advantage of proven minimal makespans.[17]

Two types of instances have been created which are different in nature. Instances of the first type comprise many jobs consisting of a small number of operations. We refer to this set of instances as 'short-jobs'. Instances of the second type comprise fewer jobs but with a larger number of operations per job. We refer to this set of instances as 'long-jobs'. All instances have a minimal makespan of 600,000 (600k), which roughly constitutes one week in seconds.

We tested the three presented encodings (basic, SEQ, MTWR) on the benchmark instances on a 3.2 GHz machine and allowed a maximum of 3600 s before a time-out occurred. For the basic encoding the built-in default variable ordering heuristic *smallestDomain* (= most constrained variable first) was employed. Each of the encodings was tested with different maximal allowed makespans: 600,000 (600k) = the optimum, 800,000 (800k) = 33.3% off the optimum, 1,000,000 (1000k) = 66.6% off the optimum, and 1,200,000 (1200k) = double the optimum. Table 2 summarizes the results for the different encodings on the benchmark problems allowing different maximal makespans. All times are given in seconds.

The performance of the basic encoding on the long-jobs benchmark is two-fold. When max allowed makespan is equal to the optimum, a schedule could be computed for all but one instance. In these cases the search problem is highly constrained. Consequently constraint propagation is very effective such that almost no search is needed. When increasing the max allowed makespan the instances become less constrained such that more search is needed. Having a max allowed makespan of 800k, only a single instance could be solved. Relaxing the max allowed makespan even more makes it easier again to come up with a schedule. Still for many instances it was not possible to produce schedules which are even double the optimum. On the

[17]Used benchmark instances, encodings and result data sets can be downloaded at http://isbi.aau.at/hint/misc.

short-jobs instances the performance of the basic encoding is even worse as those instances are less constrained.

The performance of the dispatching rule encodings is much better in general. Similarly as in [16], the MTWR rule produces overall the best results. However, the very simple SEQ rule performs quite well and sometimes even better than MTWR. Similarly as for the basic encoding, even optimal solutions can be achieved if instances are tightly constrained. In our evaluation this is the case for the long-jobs instances in combination with setting the maximum allowed makespan to the real optimum. For these cases SEQ was always able to reproduce an optimal solution. Of course, in real world domains the optimum is not known in general and also the structure of the problem instances (e.g. long-jobs vs. short-jobs) often changes depending on the customers' orders and the current product portfolio. Consequently, pursuing only a single strategy might not be indicated. Instead, a combination of different dispatching rule based encodings and increasing maximum makespans should be used.

6 Conclusions and Future Work

Elaborated engineering and sophisticated general problem-independent heuristics have significantly improved the runtime performance of general problem-solvers. However, it is a well known observation that general problem-solvers which are applied to $\mathcal{NP}$-hard problems deliver satisfiable performance up to a certain size of the problem instances. For many problems special heuristic algorithms were developed which resulted in exceptional runtime improvements compared to state-of-the-art general problem-solvers. Though, general problem solvers, in particular declarative problem solving approaches, provide superior knowledge representation features which allow the efficient development and maintenance of problem descriptions.

The main contribution of this article is to show how problem-dependent heuristics can be realized within a constraint answer set programming framework (CASP). To this end, we first introduced the novel CASP solver ASCASS (A Simple Constraint Answer Set Solver). ASCASS allows the declarative formulation of problem-specific heuristics. This is done by the usage of ASP for generating problem-dependent heuristics for CSP including variable and value selection as well as pruning strategies.

We then exemplified the realization of problem-dependent heuristics for two highly important manufacturing problem domains: the configuration of products and services and the scheduling of jobs in large-scale job shops.

In the configuration domain we focused on the real-world Partner Units Problem (PUP), which constitutes one of the hardest benchmark problems of the ASP competitions. By providing an encoding in ASCASS that includes both the declarative description of the problem and an effective heuristic for solving the problem, we could show that the non-trivial problem-dependent QuickPup heuristic can be expressed succinctly in ASCASS. Due to this heuristic, which cannot be expressed by any other ASP or CASP system, ASCASS clearly outperforms state-of-the-art ASP or CASP systems on the tested PUP instances.

For the job scheduling domain, it can be summarized that the CASP approach can successfully be applied also for large-scale job shop scheduling problem instances and the concept of dispatching rules can be easily integrated in CASP. We demonstrated the efficacy of dispatching based CASP encodings with respect to a new large-scale benchmark with proven optima and comprising up to 10000 job operations to be scheduled on up to 100 machines.

Future work will concentrate on automatic generation of heuristics that consist of CSP variable/value ordering and pruning strategies in the language of ASCASS. A second interesting research direction connecting past research of the authors is to investigate the possibilities of CASP in the area of knowledge-based recommender systems [9, 29–32].

Acknowledgements Work has been conducted in the scope of the project *Heuristic Intelligence (HINT, FFG-PNr.: 840242).*

References

1. Aschinger, M., Drescher, C., Friedrich, G., Gottlob, G., Jeavons, P., Ryabokon, A., Thorstensen, E.: Optimization methods for the partner units problem. In: CPAIOR'11, pp. 4–19. Springer, Berlin (2011)
2. Balduccini, M.: Representing constraint satisfaction problems in answer set programming. In: ICLP09 Workshop on Answer Set Programming and Other Computing Paradigms (ASPOCP'09) (2009)
3. Balduccini, M.: Industrial-size scheduling with asp+cp. In: Proceedings of the 11th International Conference on Logic Programming and Nonmonotonic Reasoning, LPNMR'11, pp. 284–296. Springer, Berlin (2011). http://dl.acm.org/citation.cfm?id=2010192.2010229
4. Bartk, R., Salido, M., Rossi, F.: New trends in constraint satisfaction, planning, and scheduling: a survey. Knowl. Eng. Rev. **25**(3), 249–279 (2010). doi:10.1017/S0269888910000202
5. Blazewicz, J., Ecker, K., Pesch, E., Schmidt, G., Weglarz, J.: Handbook on Scheduling: Models and Methods for Advanced Planning (International Handbooks on Information Systems). Springer, New York (2007)
6. Drescher, C.: The partner units problem: a constraint programming case study. In: ICTAI'12 (2012)
7. Falkner, A., Haselboeck, A., Schenner, G., Schreiner, H.: Modeling and solving technical product configuration problems. AI EDAM 115–129 (2011)
8. Felfernig, A., Friedrich, G., Jannach, D.: Conceptual modeling for configuration of mass-customizable products. Artif. Intell. Eng0 **15**(2), 165–176 (2001). http://www.sciencedirect.com/science/article/pii/S0954181001000164
9. Felfernig, A., Mairitsch, M., Mandl, M., Schubert, M., Teppan, E.: Utility-based repair of inconsistent requirements. In: Proceedings of the 22nd International Conference on Industrial, Engineering and Other Applications of Applied Intelligent Systems: Next-Generation Applied Intelligence, IEA/AIE '09, pp. 162–171. Springer, Berlin (2009)
10. Garey, M.R., Johnson, D.S.: Computers and Intractability; A Guide to the Theory of NP-Completeness. W. H. Freeman & Co., New York (1990)
11. Gebser, M., Kaminski, R., Kaufmann, B., Schaub, T.: Answer Set Solving in Practice. In: Synthesis Lectures on Artificial Intelligence and Machine Learning. Morgan and Claypool Publishers (2012)

12. Gebser, M., Kaufmann, B., Romero, J., Otero, R., Schaub, T., Wanko, P.: Domain-specific heuristics in answer set programming. In: 27th AAAI Conference (AAAI'13), pp. 350–356 (2013)
13. Gelfond, M., Lifschitz, V.: The stable model semantics for logic programming. In: Kowalski, R., Bowen, K. (eds.) Proceedings of the Fifth International Conference and Symposium of Logic Programming (ICLP'88), pp. 1070 – 1080. MIT Press (1988)
14. Harvey, W.D., Ginsberg, M.L.: Limited discrepancy search. In: Proceedings of the 13th International Joint Conference on Artificial Intelligence, pp. 607–613. Morgan Kaufmann (1995)
15. Hildebrandt, T., Goswami, D., Freitag, M.: Large-scale simulation-based optimization of semiconductor dispatching rules. In: Proceedings of the 2014 Winter Simulation Conference, WSC '14, pp. 2580–2590. IEEE Press, Piscataway, NJ, USA (2014). http://dl.acm.org/citation.cfm?id=2693848.2694175
16. Kaban, A.K., Othman, Z., Rohmah, D.S.: Comparison of dispatching rules in job-shop scheduling problem using simulation: a case study. Int. J. Simul. Model. **11**(3), 129–140 (2012). doi:10.2507/IJSIMM11(3)2.201
17. Lewis, M.D.T., Schubert, T., Becker, B.W.: Speedup techniques utilized in modern sat solvers. In: Proceedings of the 8th International Conference on Theory and Applications of Satisfiability Testing, SAT'05, pp. 437–443. Springer, Berlin (2005). doi:10.1007/11499107_36
18. Lierler, Y., Smith, S., Truszczynski, M., Westlund, A.: Weighted-sequence problem: ASP vs CASP and declarative vs problem-oriented solving. In: Proceedings of the 14th International Conference on Practical Aspects of Declarative Languages, PADL'12, pp. 63–77. Springer, Berlin (2012). doi:10.1007/978-3-642-27694-1_6
19. Mellarkod, V.S., Gelfond, M., Zhang, Y.: Integrating answer set programming and constraint logic programming. Ann. Math. Artif. Intell. **53**(1–4), 251–287 (2008). doi:10.1007/s10472-009-9116-y
20. Mittal, S., Frayman, F.: Towards a generic model of configuration tasks. In: 11th International Joint Conference on AI—IJCAI'89, vol. 2, pp. 1395–1401. Morgan Kaufmann Publishers Inc., San Francisco, CA, USA (1989)
21. Ostrowski, M., Schaub, T.: ASP modulo CSP: the clingcon system. Theory Pract. Log. Program. **12**(4–5), 485–503 (2012). doi:10.1017/S1471068412000142
22. Sadegheih, A.: Scheduling problem using genetic algorithm, simulated annealing and the effects of parameter values on GA performance. Appl. Math. Model. **30**(2), 147–154 (2006). doi:10.1016/j.apm.2005.03.017. http://www.sciencedirect.com/science/article/pii/S0307904X05000521
23. Sadeh, N.M., Fox, M.S.: Variable and value ordering heuristics for the job shop scheduling constraint satisfaction problem. Artif. Intell. **86**, 1–41 (1996)
24. Sebastiani, R.: Lazy satisfiability modulo theories. J. Satisfiability, Boolean Model. Comput. **3**, 141–224 (2007)
25. Stump, B., Badurdeen, F.: Integrating lean and other strategies for mass customization manufacturing: a case study. J. Intell. Manuf. **23**(1), 109–124 (2012). doi:10.1007/s10845-009-0289-3
26. Stumptner, M.: An overview of knowledge-based configuration. AI Commun. **10**, 111–125 (1997). http://dl.acm.org/citation.cfm?id=1216064.1216069
27. Syrjänen, T.: Cardinality constraint programs. In: JELIA 2004, Lecture Notes in Computer Science, vol. 3229, pp. 187–199. Springer (2004)
28. Teppan, E.C.: Re-configuring legacy instances of the partner units problem. In: International Conference on Tools with Artificial Intelligence (ICTAI'12), pp. 154–161. IEEE (2012)
29. Teppan, E.C., Felfernig, A.: Asymmetric dominance-and compromise effects in the financial services domain. In: 2009 IEEE Conference on Commerce and Enterprise Computing, pp. 57–64 (2009). doi:10.1109/CEC.2009.69
30. Teppan, E.C., Felfernig, A.: Calculating decoy items in utility-based recommendation. In: Proceedings of the 22nd International Conference on Industrial, Engineering and Other Applications of Applied Intelligent Systems: Next-Generation Applied Intelligence, IEA/AIE '09, pp. 183–192. Springer, Berlin (2009)

31. Teppan, E.C., Felfernig, A.: Impacts of decoy elements on result set evaluations in knowledge-based recommendation. Int. J. Adv. Intell. Paradigms **1**(3), 358–373 (2009). doi:10.1504/IJAIP.2009.026573
32. Teppan, E.C., Felfernig, A.: Minimization of decoy effects in recommender result sets. Web Intelli. Agent Sys. **10**(4), 385–395 (2012)
33. Teppan, E.C., Friedrich, G., Falkner, A.: Quickpup: a heuristic backtracking algorithm for the partner units configuration problem. In: International Conference on Innovative Applications of AI (IAAI'12), pp. 2329–2334. AAAI (2012)
34. Teppan, E.C., Friedrich, G., Gottlob, G.: Tractability frontiers of the partner units configuration problem. J. Comput. Syst. Sci. (2016). doi:10.1016/j.jcss.2015.12.004. http://www.sciencedirect.com/science/article/pii/S0022000016000167
35. Watson, J.P., Beck, J.C., Howe, A.E., Whitley, L.D.: Problem difficulty for tabu search in job-shop scheduling. Artif. Intell. **143**(2), 189–217 (2003). doi:10.1016/S0004-3702(02)00363-6

Printed in the United States
By Bookmasters